General Biology
BI 101
Laboratory Manual

Third Edition
Montgomery College

Edited by:
Cyrus MacFoy
Nelson Bennett

Kendall Hunt
publishing company

Kendall Hunt
publishing company

www.kendallhunt.com
Send all inquiries to:
4050 Westmark Drive
Dubuque, IA 52004-1840

Copyright © 2005, 2007, 2012 by Kendall Hunt Publishing Company

ISBN 978-1-4652-0819-4

Printed in the United States of America
10 9 8 7 6 5 4 3 2 1

Contents

Introduction

Welcome

General biology is meant to provide you with an overview of one small part of human knowledge, the study of living things. This encompasses a wide range of ideas ranging from the functions of your body to the relationships between small things living in a mud puddle. Where possible an international perspective is emphasized. In the laboratory portion of this course you have the opportunity of working in the biology lab and getting some first-hand experience. The labs have been selected to accompany the lectures in the course, which will give you a chance to "do something" with the ideas you are learning. You will find that the lab helps you learn biological ideas, and, hopefully, can be both fun and interesting.

Working in the Lab Room

Like a kitchen in a house, the "laboratory" is a room where work is done. If you have ever tried cooking in the kitchen, then you know that it is important to be familiar with the locations of ingredients and tools, how the tools work, and following the instructions in the recipe. On your first day of lab you will review the layout of the lab room, its safety features, and lab safety rules. Every lab meeting will require you to use either some different equipment or different "ingredients" as described in the lab instructions in this manual (your recipe book). Additional details, changes, and safety tips may be reviewed by your instructor.

If you wish to have a successful lab session, then you need to make sure that you carefully read the lab instructions for that day before you come to class. Just as you cannot expect to cook a special dish if you forget the ingredients, you cannot expect to fully participate in lab if you have not prepared. In lab, you work together with a lab partner or in lab groups, and you become a burden to everyone, not just yourself, if you are not prepared to do your part.

Etiquette

In addition to being prepared, you need to make sure that you are mindful of what you are doing and how it may affect others in class. Lab time is a time to do work. Although you may have fun working in groups or with your partner, you should keep some things in mind:

- Maintain a safe, studious atmosphere for yourself and others by not talking too loud or horsing around.
- Keep your work area "clean" by having only what you need on the table. Place other items below the table, and clean up after yourself.
- Keep the room neat by returning specimens and equipment to their appropriate places when you are finished with them.

- Be prepared, by reading the lab twice before class. You will be able to make the most of your time. Homework should be done at home!

- Work together as a team! Working together will help you get things accomplished more efficiently, but you also need to make sure that your lab group/partner understands what is being done and why.

- For health and cleanliness reasons, eating, drinking, smoking, storing food, chewing gum and applying cosmetics are not permitted in the laboratory. You may bring food and beverage containers with you into the lab as long as they stay sealed and stored in the black bookcase by the door.

- Open toed shoes are not allowed in the laboratory.

- Your lab bench area must be decontaminated using a commercially prepared disinfectant at the end of every laboratory period and after ever spill. You are welcome to further protect yourself by decontaminating your lab bench area at the beginning of lab.

- Disposable materials such as gloves, mouthpieces, swabs, and toothpicks that come into contract with body fluids should be placed in a disposable autoclave bag for autoclaving. Any equipment such as glassware contaminated by body fluids must be placed in a disposable autoclave bag for autoclaving or placed directly into a 10% bleach solution before reuse or disposal.

- Disposable gloves are recommended for dissection of preserved materials.

- Report all spills or accidents, no matter how minor, to your instructor.

- Students who are pregnant, taking immunosuppressive drugs, or have any other medical condition that might necessitate special precautions in the laboratory must inform your instructor immediately.

- Never work alone in the laboratory.

- Do not allow water or any solution to come into contact with electrical cords or outlets. Make sure your hands are dry when you handle electrical connectors. If electrical equipment crackles, snaps, or begins to smoke, do not attempt to disconnect it. Inform your instructor immediately.

- Do not touch broken glassware with your hands. Use a broom and dustpan. Place broken glassware in the container marked for that purpose.

- Let your instructor know if you are colorblind, as many procedures require discrimination of colors.

- Children are not permitted in the laboratory. Adult guests are allowed only with the permission of the instructor.

- Push in stools and chairs at end of lab to avoid tripping hazard.

- Wash hands before leaving the laboratory.

- Be safe, and let's have a great semester together!

Measurement

Establishing a value system based on a scale of measurement gives people a means to compare and communicate when dealing with the exchange of materials. Measuring units were developed to simplify and to standardize the selling or exchange of materials. Definitions or standards were needed for units of measure. How much is a "bag" of potatoes? How much milk is in a "bottle" of milk, etc.?

Most Americans are familiar with the English System of Measurement. This system uses inches, feet, and yards for length measurements. Cups, pints, quarts, and gallons are used for volume measurements. Ounces, pounds, and tons are used for weight measurements. However, the English system is cumbersome. There is no consistency between converting units e.g., 16 ounces = 1 pound, 12 inches = 1 foot, 5,260 feet = 1 mile. This requires the memorization of many conversions. Also, in different countries the same term may have different values.

In an attempt to standardize measurement worldwide the International Metric System was developed. This system provides precision by powers of 10. There is uniformity in measurements of length, volume, and weight, and the same prefixes are used in all of the different forms of measurement. For example, milli always means 1/1000.

The Metric System has the following 3 standard units.

Length	=	the meter (m)
Volume	=	the liter (L)
Mass	=	the gram (g)

The units above are combined with the following prefixes.

micro (μ)	=	1/1,000,000
milli (m)	=	1/1,000
centi (c)	=	1/100
deci (d)	=	1/10
deka (da)	=	10
hecto (h)	=	100
kilo (k)	=	1000

You should memorize the above terms and symbols.

"Measurement" from *Introductory Biology Lab Manual* by Jim Goodwin, Mark Barnby and Carol Dixon.
Copyright © 2000 by James Goodwin, Mark Barnby and Carol Dixon. Reprinted by permission of the authors.

Directions

Complete the problems on the first page of your Data/Answer Sheet after reading each of the following subsections.

Conversion in the Metric System

Conversion from one level of the metric system to another level is very basic, the difference between each level is always an increase or decrease of 10. Because of this fact, conversion simply involves moving the decimal point to the right or to the left, without actually changing the number values. The scale below should make conversions very easy.

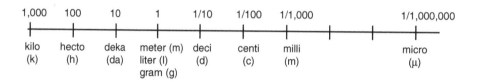

To convert any measurement, simply find its unit on the scale and then count how many places and what direction (right or left) you are moving on the scale. You will move the decimal point exactly the same as you moved on the chart. For example, if you want to convert 121 cm (centimeters) to km (kilometers), you go to the chart and count how many places and what direction you will move from c to k, which is 5 places to the left. Now take 121 and move the decimal 5 places to the left to get 0.00121. Therefore, there is 0.00121 km in 121 cm. How many μm are in 121 cm?

Metric to English Conversion

Here are five metric to English conversions you should know.

2.54 cm	=	1 in.
1 m	=	1.1 yds.
455 g	=	1 lb.
1 kg	=	2.2 lbs.
3.8 L	=	1 gallon

Fractions, Decimals, and Percentages

Parts of a whole are usually measured in one of two ways.

    1. Fractions 1/5, 1/4, 1/2

    2. Decimals 0.2, 0.25, 0.5

A. Fractions can be converted to decimals by dividing the top number of the fraction (numerator) by the bottom number of the fraction (denominator).

B. Decimals can be converted to fractions by putting the decimal number over 1, and then reducing the fraction to its lowest form. For example 0.24 = 0.24/1 = 24/100 = 6/25.

C. Fractions are converted into percentages by first converting the fraction into a decimal, and second, by multiplying the decimal by 100.

D. Decimals can be converted to percentages by multiplying the decimal by 100.

Practice in Metric Measurement

A. Use the appropriate devices, (as demonstrated by the instructor), to measure the length (ruler), mass (balance or scale), or volume (pipette, graduated cylinder, volumetric flask) of samples provided by the instructor. You may need to express the measurement in *scientific notation*. Scientific notation is written as a product of a decimal number between 1 and 9 and the number 10 raised to the proper power. For example 450,002 can be written as 4.50002×10^5; 0.000567 can be written 5.67×10^{-4}.

B. When a measurement is reported, the measurement should contain the certain digit(s) and only one estimated (uncertain) digit. All certain digits plus *one* uncertain digit (estimated) are known as *significant figures*.

Procedure

    1. Use a meter stick to measure the width of the table in cm and convert to mm, m, and km.

    2. Use a balance and weigh an object provided by the instructor. Record its mass in g and convert to mg, µg, and kg.

    3. Use a pipette to transfer 10 ml of water into a 10 ml graduated cylinder. Record the position of the meniscus.

    4. Use a 250 ml Erlenmeyer flask to transfer 100 ml of water into a 100 ml graduated cylinder. Record the position of the meniscus.

C. Use a thermometer to measure the temperature of boiling water, an ice bath, and water at room temperature. Record the values in degrees Celsius and convert them into degrees Fahrenheit.

    $C° = 5/9 \ (F°-32)$ and $F° = (9/5 \ C°) +32$

Preparation of Solutions and Dilutions

Solutions are homogenous mixtures of a *solute,* the substance that is dissolved, and a *solvent,* the dissolving liquid. In biology, two of the more common types of solutions are percent solutions and molar solutions.

A percent solution is the number of grams of solute per 100 ml of solvent (% w/v). When mixing liquids, a percent solution is the number of ml of a solution in a combined total of 100 ml of solvent.

A molar solution is the number of moles (a mole of a substance is equal to its MW in grams) in a total of 1 L of solvent.

Use the following formulas to calculate the amount of substance needed to prepare a solution at a particular concentration and volume, and to dilute a stock solution. C = concentration & V = volume; $_i$ = initial & $_f$ = final.

$$CV = \text{amount}$$

$$C_iV_i = C_fV_f$$

Procedure

1. How would you prepare 200 ml of a 1.5% solution of NaCl in water. Calculate the amount of NaCl, and the volume of H_2O you would use.

2. How would you now use the above solution to prepare a 50 ml solution of 0.075% NaCl. Record the C_iV_i, C_fV_f on page 7.

Data/Answer Sheet 1

Metric Conversions

1. 1 cm   = _____ m
2. 2120 mm = _____ m
3. 12 m   = _____ mm
4. 9 dm   = _____ m
5. 14 m   = _____ hm
6. 1 m   = _____ km
7. 1 km   = _____ m

8. 1000 ml   = _____ L
9. 21 L   = _____ cl
10. 10 dal   = _____ L
11. 8 kl   = _____ L
12. 1 kg   = _____ g
13. 2421 g   = _____ kg
14. 683 mg   = _____ g
15. 3.2 cm   = _____ μm

Metric to English Conversions

1. 10 inches = _____ cm
2. 31 lbs.   = _____ grams
3. 50 liters = _____ gallons

4. 2730 g   = _____ lbs.
5. 15.2 gal   = _____ liters
6. 10 meters = _____ yards

Fractions, Decimals, and Percentages

Change the following fractions into decimals.

7/10_____   3/4_____   2/5_____

4/9_____   3/10_____   1/3_____

11/25_____   9/32_____   7/8_____

Change the following decimals to fractions.

0.23_____   0.345_____   0.55_____

0.44_____   0.5_____   0.789_____

0.67_____   0.98_____   0.84_____

Convert the following decimals to percentages (%).

0.64   = _____%   0.3465 = _____%   0.888   = _____%

0.89   = _____%   0.987   = _____%   0.99   = _____%

0.77   = _____%   0.5432 = _____%   1.45   = _____%

Name: _____   Date: _____

Data/Answer Sheet 2

Practice in Metric Measurement

1. Length measurement: width of table

   _____ cm

   _____ mm

   _____ m

   _____ km

   | Rewrite in Scientific Notation |

   _____ cm

   _____ mm

   _____ m

   _____ km

2. Mass Measurement:

   _____ g

   _____ mg

   _____ μg

   _____ kg

   | Rewrite in Scientific Notation |

   _____ g

   _____ mg

   _____ μg

   _____ kg

3. Temperature Measurement:

   _____ C° boiling water

   _____ C° ice bath

   _____ C° room temp.

   | Convert to F° |

   _____ F°

   _____ F°

   _____ F°

Preparation of Solutions and Dilutions

1. 1.5% NaCl solution

   _____ g NaCl needed

   _____ ml H_2O needed

2. 50 ml of 0.075% NaCl solution

   Ci = _____ C_f = _____

   V_i = _____ V_f = _____

   How much of 1.5% NaCl used   _____ ml

   How much water added   _____ ml

2

The Scientific Method Introduction to Research

The Scientific Method of Inquiry

This is the tool that scientists use to solve problems in the laboratory and in the field. Incidentally we all use it every day as well, to solve our daily problems.

The basic steps are as follows:

1. **Observation**—Involves making careful observation of nature either directly or indirectly, using the senses e.g. sight, sound, smell, touch, taste and instruments to collect information.

2. **Question**—Define the problem by asking a question based on your observation.

3. **Hypothesis**—Formulate a hypothesis. This is a tentative answer to the question—an educated guess—based on the observed data, your knowledge and past experience.

4. **Prediction**—A prediction is made based on the hypothesis; it is a logical conclusion to be expected in view of the hypothesis; usually expressed as follows: if the hypothesis is correct then a particular outcome is expected.

5. **Experiments**—Predictions are tested by observations and/or experimentation. The experiment consists of a control group and an experimental group. There are three variables viz. **the independent variable** i.e. the condition that you are evaluating for its effect on the dependent variable; it is the one that you manipulate. The **dependent variable** is the variable that is measured and the **controlled variables** are all the other conditions, which are kept constant throughout the experiment. In the experiment the **experimental group** is exposed to the independent variable but the **control group** is not.

6. **Results**—These are collected and analyzed.

7. **Conclusion**—Conclusions about whether the hypothesis is accepted or rejected is made, based on the results of the experiment.

Exercise 1

Introduction to the scientific method

You have a problem e.g. on the way to your car after a biology lesson you discovered that you cannot find your car keys. There are several possibilities about what may have happened to it. One has already been done for you. State any two other hypotheses.

Hypothesis 1: You locked the keys in your car

Hypothesis 2:

Hypothesis 3:

After stating a hypothesis you must test it. How can you test each of these three hypotheses?

1.

2.

3.

Exercise 2

Forming a hypothesis for your own experiment and testing your hypothesis

Form a hypothesis using the examples from this list or from your own.

1. Body weight; height; pulse rate; breathing rate; head circumference; arm length.

   e.g. the taller you are the faster your heart rate

Available for your use are: scales; tape measures; meter sticks, and stop watch.

How are you going to get the data you need to test your hypothesis?

Write the steps you will follow to perform your experiment.

Ask your classmates (at least 10) to serve as test subjects (guinea pigs).

Perform your experiments and record your results in a table.

Rearrange these results from smallest to largest or from shortest to tallest or lightest to heaviest as the case may be.

Is there any relationship between the two factors? Did the data support your hypothesis?

Draw a graph (line or bar) to show the relationship.

Discuss your results and state your conclusion, which should be based on the hypothesis and the data you collected. The conclusion should be supported by data. Evaluate your experimental design; include any reflections. Use the format on page 12 to write a detailed report.

Exercise 3

Practice Graphing Experimental Data

Most experiments are designed to establish a cause and effect relationship between 2 variables: the INDEPENDENT VARIABLE is the factor whose effect is tested in the experiment. Its magnitude is purposely varied by the investigator. The DEPENDENT VARIABLE is the factor whose value is measured by the investigator. It's magnitude depends on the magnitude of the independent variable. All other variables are held constant for the duration of the experiment and are referred to as CONTROLLED VARIABLES.

Experimental data is graphed in order to display the relationship between the independent and the dependent variable. The value of the INDEPENDENT VARIABLE is plotted on the horizontal axis (x-axis) of the graph. The value of the DEPENDENT VARIABLE is plotted on the vertical axis (y-axis) of the graph.

During the course of the semester, you will be asked to construct 2 types of graphs.

1. LINE GRAPHS are used when both the independent and dependent variables are measured over a continuous scale. Use the following check list in preparing a line graph.

   a. Use a ruler to construct the x and y axes. Place the independent variable on the x-axis and the dependent variable on the y-axis.

   b. Label each axis with quantity and units; for example, the x-axis might be labeled Time (sec.), while the y-axis is labeled Glucose Concentration (g/l).

   c. Select a scale for each axis that allows you to plot the largest value measured. Then, mark off uniform intervals. Unless otherwise labeled, the origin is understood to be (0,0).

   d. Use a pencil to plot the data set.

   e. Unless otherwise instructed, draw the smooth curve or single straight line determined by the data points.

   f. Give the graph a descriptive title.

2. BAR GRAPHS are used when the independent variable represents separate or discrete categories or groups. The same rules apply, except that:

   a. Categories are located along the x-axis.

b.   A vertical bar is drawn to represent the magnitude of the dependent variable for each category.

Following the guidelines given above, prepare a graph of each of these data sets interpret each graph.

Growth of a Coleus Plant	
Day	**Height (cm)**
8	5
14	10
24	15
30	17.5
33	18

Protein Content of Foods	
Food	**% Protein**
Rice	7
Beans	8
Peanuts	27
Milk	3
Fish	27
Beef	23
Chicken	33

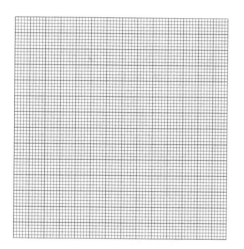

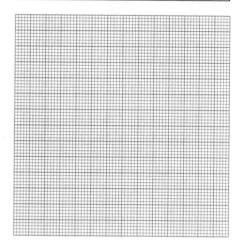

Writing your report of Exercise 2—use the following format:

Title—Use a suitable title for the report.

Introduction—This should include background information and observations required for the reader to understand the experiment. Describe the observations that caused you to identify the problem for investigation; also state the problem you defined based on your observations; state the hypothesis and briefly describe how you tested it.

Materials—Describe the materials and equipments you used.

Methods—Describe the procedures you used to collect and analyze the data.

Results—Tabulate your data and graph your results appropriately with the X-axis containing the independent variable and the Y-axis the dependent variable; give the graph a title and a key. Interpret the graph.

Discussion—Discuss your results e.g. does your data support the hypothesis; did you experience any difficulties?

Conclusion—What is your conclusion?

References/Bibliography—List appropriately all articles, books etc, consulted for this experiment.

Taxonomy: Construction and Use of Taxonomic Keys

Objectives

When you have completed this topic you should be able to:

1. List the major taxa in the correct order from most general to least general.

2. Use a simple dichotomous key to classify organisms.

3. Construct a simple dichotomous key.

Introduction

Organisms are grouped into categories on the basis of their evolutionary relationships and modern day similarities. The science of classifying organisms and placing them into different categories is called <u>**taxonomy**</u> or <u>**systematic biology**</u>.

1. Major Category Levels (taxa)

Seven major category levels are commonly used to classify organisms into related groups. Each category level is called a **taxon** (plural = taxa). The seven principal taxa in descending order, from most general to least general are:

kingdom, phylum (formerly called a division in the kingdom plantae), class, order, family, genus, species

A **kingdom** is a very large grouping (i.e., the most general of the taxa listed) which includes a large variety of different types of organisms that share a few important characteristics that distinguish them from members of the other kingdoms. The least general, or most specific, category level is the **species** level, which consists of a single type of organism whose members interbreed in natural populations. The **genus** (or generic level) level is a group of similar, closely related species that normally do not interbreed in natural populations.

2. The Domain Taxon (also see lab topic 2)

A more recent system of classification includes a taxon above the kingdom level. This taxon is called a **domain,** and is a kind of "super category." In the **domain classification system,** there are three domains (Bacteria, Archaea, and Eukarya), and the domains are divided into kingdoms. The idea behind the three domain system is that the domains represent three major branches of life.

3. Current Classification Systems (also see lab topic 2)

There is nothing absolute about a classification system. Instead, as we learn more about the organisms we already know (e.g., by analyzing their DNA), discover organisms that we have never seen before, or find new evidence in the fossil record, the classification systems may be revised. For example, until recently, the classification system that was used was a **five kingdom classification system.** However, as scientists learned more about a group of bacteria-like organisms (referred to as Archaea or Archaebacteria) that previously had been included with "ordinary" bacteria in the same Kingdom, they decided that they were too different to remain grouped in the same Kingdom.

Although just about all scientists now agree that these "special" bacteria-like organisms should not be included in the same Kingdom with "ordinary" bacteria, scientists are not yet in agreement about the kind of classification system that should replace the old system. Of course, this means that there are currently two systems of classification that have to be learned. These systems are the Six Kingdom Classification System and the Three Domain Classification System. The figures below show how the Three Domain System and the Six Kingdom System relate to each other.

Three Domain System

Domain **Bacteria**	Domain **Archaea**	Domain **Eukarya** (includes the Kingdoms Protista, Plantae, Fungi, and Animalia)

Six Kingdom System

Kingdom **Eubacteria** (Bacteria)	Kingdom **Archaebacteria** (Archaea)	Kingdom **Protista**	Kingdom **Plantae**	Kingdom **Fungi**	Kingdom **Animalia**

4. Species Names (binomial system of nomenclature)

All organisms have a scientific species name which consists of two parts. Because there are always two parts to a species name, this kind of naming is referred to as a **binomial system of nomenclature.** The first part of the species name is the **genus** to which the organism belongs. The second part of the species name is called the **specific epithet** (or **specific name**), which, when combined with the genus (or **generic name**) name, refers to a particular species within the genus. For example, the species name for dogs is *Canis familiaris*. The term "*Canis*" is the genus (or **generic name**) to which dogs belong. The term "*familiaris*" is the specific epithet, which when combined with "*Canis*" constitutes the species name for dogs. The genus "*Canis*" includes other types of organisms. For example, wolves are classified in the genus "*Canis.*" The term "*lupus*" is the specific epithet, which when combined with "*Canis*" constitutes the species name for wolves (i.e., *Canis lupus*). When writing species names, the generic name is capitalized and the specific epithet is in the lower case, and both terms must be underlined *or* typed in italics.

5. Examples of Classifications

Humans (animal) are classified as follows:

DOMAIN . . . . Eukarya

   KINGDOM . . . . Animalia   more general

      PHYLUM . . . . Chordata

         CLASS . . . . Mammalia

           ORDER . . . . Primates

              FAMILY . . . . Hominidae

                 GENUS (or generic name) . . . *Homo*

                   SPECIES . . . . *Homo sapiens* more specific

The coast redwood (plant) is classified as follows:

DOMAIN . . . . Eukarya

   KINGDOM . . . . Plantae

      DIVISION . . . . Coniferophyta

         CLASS . . . . Coniferopsida

           ORDER . . . . Coniferales

              FAMILY . . . . Taxodiaceae

                 GENUS (or generic name) . . . Sequoia

                   SPECIES . . . . *Sequoia sempervirens*

Taxonomic Keys

1. How to Construct a Dichotomous Key

Once organisms are described, classified on the basis of their distinguishing characteristics and evolutionary relationships, and given a species name, it is possible to devise a system that allows others to readily identify an organism that they have never seen before. The most common system used for identifying organisms is called a **taxonomic key,** and the most common type of taxonomic key is called a **dichotomous key.** In a **dichotomous key,** a series of contrasting characteristics are used to lead the investigator, by a process of elimination, to the identity of a specific organism.

In constructing a dichotomous key, it is useful to first describe the contrasting characteristics of the organisms that the key will be used to identify. This technique is illustrated in Table 3.1 for four different species of oaks. The next step is to arrange the contrasting characteristics in pairs, presenting the investigator with a choice that will refer the investigator to another set of contrasting characteristics, until the organism is identified. This is illustrated in the "Taxonomic Key to Selected Oaks" provided below Table 3.1.

Table 3.1: Diagnostic Table for Some of the Oaks

	Lobe shape	**# of lobes**	**Lobe depth**	**Acorn length**	**Bark texture**
Quercus rubra	Pointed lobes	7–11	Less than half-way to the midrib	2.25–2.5 cm	Wide ridges
Quercus velutina	Pointed lobes	7	Less than half-way to the midrib	1.5–2.0 cm	Narrow ridges
Quercus palustris	Pointed lobes	5–7	More than half-way to the midrib	1.0–1.3 cm	Smooth (no ridges)
Quercus imbricaria	Unlobed	0	Unlobed	6.0–8.0 mm	Smooth or with ridges

2. How to Use a Dichotomous Key

If you were walking across the campus and came upon an oak tree which you could not identify (let's say it is a pin oak, which is the common name for *Quercus palustris*), you could use the key to oaks given above as follows:

You would first examine the leaves of the tree to determine whether they were lobed or unlobed (the first pair of questions). Since the leaves of a pin oak are lobed, you would select the statement numbered "1a" (leaves lobed) and be referred to the second couplet, which asks that you measure the length of acorns produced by the tree. When measuring the acorns, you would find that they were less than 2.0 cm. Therefore, you would select the second alternative (2b) and be referred to the third couplet that asks you to examine the bark of the tree. Upon examining the bark, you would find that the bark was smooth and without ridges. This leads you to accept the second statement (3b) in couplet 3, which gives you the identity of the tree as a pin oak *(Quercus palustris)*. By following this key, you have learned that only one of the four oaks described in the key, the pin oak, has all of the following characteristics: it is an oak tree with lobed leaves, acorns that are less than 2.0 cm in length, and a smooth bark without ridges.

"Taxonomic Key to Selected Red Oaks"

1a. Leaves lobed ⟶ **go to 2**

1b Leaves unlobed ⟶ *Quercus imbricaria*

2a Acorns 2.25–2.5 cm in length ⟶ *Quercus rubra*

2b Acorns less than 2.0 cm in length ⟶ **go to 3**

3a Bark with distinct narrow ridges ⟶ *Quercus velutina*

3b Bark is smooth and without ridges ⟶ ***Quercus palustris***

Lab Exercises:

In today's lab, there are three exercises dealing with keys.

Exercise #1: Constructing a Key to Fruits and/or Vegetables

You will be given a tray containing a variety of fruits and/or vegetables and should construct a dichotomous key. The following is an example of how a dichotomous key to carrots, oranges, apples, tomatoes, lemon and grapefruit might look. After reviewing this example, construct a dichotomous key to the fruits and/or vegetables in your tray.

1a the color is red ⟶ go to 2

1b the color is not red ⟶ go to 3

2a the texture is crunchy ⟶ apple

2b the texture is soft ⟶ tomato

3a the color is yellow ⟶ go to 4

3b the color is orange ⟶ go to 5

4a the shape is round ⟶ grapefruit

4b the shape is oval ⟶ lemon

5a has a thick skin ⟶ orange

5b has a thin skin ⟶ carrot

Exercise #2: Constructing a Key for a Variety of Metal Fasteners

Procedure:

In this exercise, you are asked to construct a key for an assortment of metal fasteners provided at each table. To do this, you should:

1. Look at all of fasteners and identify one characteristic that can be used to separate the fasteners into two piles. For example, you may be able to separate the fasteners on the basis of length (longer than 1 inch, shorter than 1 inch). Separate the pile into two piles according to this characteristic. Write down the pair of characteristics as 1a and 1b and write down the number 2 at the end of one of the choices as follows:

   1a   longer than 1 inch ———————➤ 2

   1b   shorter than one inch ———————➤ ?

2. Look at one of the piles, the one that you wrote the number 2 after, and leave the other pile alone for a while. Now identify a characteristic that separates this pile into two smaller piles. For example, some have heads and some don't. Then separate the pile into two smaller piles and write down this characteristic as the second pair as follows:

   2a   have heads ———————————➤

   2b   don't have heads ———————➤

   If only one of the fasteners in the group has a head, you would be able to write the name of the fastener next to line 2a. You would write the number 3 next to the line with the "have heads" characteristic, and begin to separate the have heads pile.

3. Once all of the fasteners in a sub-pile have been identified you need to go back to the sub-piles that have not been separated, putting a number next to the line with the characteristic that created the sub-pile to refer the investigator to a pair of characteristics that further separates the sub-pile. Continue until all fasteners are identified.

Exercise #3: Using a Published Key to Common Trees

The third exercise involves using a published key to Common trees. Cuttings from a number of different types of trees are available in the lab for you to identify using the key. In addition, your instructor may take you on a "field trip" to identify trees growing on campus. Note: since many of the trees on campus are not native to this region, you may come across a tree that you are unable to identify using this particular key.

OR Exercise #4: Construct a Key Using Selected Wood Samples Provided

Clean Up Procedures

Before you leave the lab today, as always, follow the specific written, clean up directions at your lab bench which include organizing your lab tray, disposing of all lab waste properly, and wiping your lab bench with germicidal solution and a *wet* sponge.

Student Observations, Notes, and Drawings

4

Microscopy

Introduction

The compound light microscope is a valuable tool for any biology student. The following discussion should allow the student to feel comfortable with the use and principles of light microscopy.

The Parts of the Compound Light Microscope

The **ocular** lenses, or eyepieces, are located at the top of the scope. Notice that the interpupillary distance can be adjusted for each user of the scope. Also notice that the left eyepiece is focusable to adjust for differences in the user's eyes. Focus the specimen through the right eyepiece while the left eyepiece is covered. Next, focus the left eyepiece by covering the right eyepiece and turning the left eyetube until the image is in clear focus. The oculars magnify the image by a factor of 10×; this is the second magnification. The first magnification is performed by the **objective** lenses. There are four objective lenses located on a **revolving nosepiece.** The first is the **scanning** lens which has a magnification of **4×.** The second objective lens is the **low** power objective with a magnification of **10×.** The third lens is the **high-dry** objective lens which has a magnification of **40×.** The fourth lens is the **oil immersion** lens with a magnification of **100×.** The **total magnification** of the image being viewed is determined by multiplying the ocular magnification by the magnification being used. Total magnification notations should be made with all drawings made during lab.

Slides are held on the scope by the use of **slide clips** seen on the surface of the **stage.** Carefully place slides in these holders; do not allow the clip to snap onto the slide as this may cause chipping of the slide. Slides being viewed are moved about the stage by the use of **vertical** and **horizontal knobs.** Do not push slides around the stage; this will strip the gears of the controls.

Beneath the stage is the **condenser.** The purpose of the condenser is to direct the light rays onto the specimen being viewed. The condenser should always be in an up position, which is just beneath the stage. The **iris diaphragm** is part of the condenser; you will see this as a black lever that can be moved in a left to right fashion. The iris diaphragm controls the amount of light entering the condenser. Less light into the condenser means less light on the specimen which will cause the specimen to have more contrast. More light leads to less contrast especially when working with very small or very thin specimens. You may have to adjust the iris diaphragm for each specimen viewed.

Focusing controls are located on the back base of the microscope. The **coarse focus** control is the larger knob located next to the base. Notice the nosepiece moves a considerable amount when using this control. The coarse focus is only used with the scanning lens. The **fine** focus control is the smaller outside knob; this is used to fine-tune the image until it is clear. The microscope is said to be parfocal which means the image stays relatively in focus when changing from one magnification to another. For this reason, when changing objective lenses you should only have to adjust the fine focus control.

Resolution is being able to see two separate objects as two distinct objects. In order to see close objects as separate objects, light must pass between them. **Resolving power** is a measure of the distance between these objects. The better the resolving power of the microscope, the less distance between the objects which in turn means the objects can be very small. The equation for resolving power is as follows:

$$RP = \frac{\textbf{wavelength}}{2 \times \textbf{NA}}$$

NA stands for **numerical aperture;** this number is usually printed on the lens. It is a measure of the cone of light entering the objective lens as seen below:

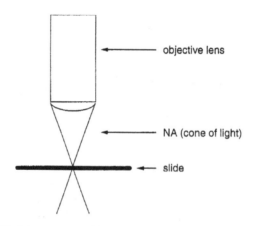

In order to view small objects a small value for RP is desired. To obtain a small number, a small value for wavelength can be divided by a larger number. For example, if 2 is divided by 4, the result is 0.5. If the wavelength used is small and is divided by a larger value for NA, a small number for RP will be obtained. Blue wavelengths of light are smaller than other light rays. Notice that there is a blue filter on top of the light source. Having the condenser closer to the stage will widen the cone of light as far as possible giving a large value for NA. Using these principles we have obtained the best resolution possible for viewing microorganisms.

Notice as higher magnification objectives are used, the working distance decreases. **Working distance** is the distance between the specimen and the objective lens. Also the opening in each objective lens becomes smaller as you increase magnification. Many light rays will be refracted away from the lens and so the image may darken as you increase magnification. Adjust the light control or the iris diaphragm as needed. Notice also that the field of view (the area being viewed on the slide) decreases as magnification increases. Make sure before increasing magnification that the area to be viewed is close to the center of the field of view.

Oil immersion is used to view extremely small specimens such as bacteria. Bacteria must be viewed under oil immersion in order to determine true shape, size, and color. Different materials have a characteristic refractive index which is the amount light bends when passing through the material. When oil immersion is used, fewer light rays are refracted as they pass through the glass slide because glass and immersion oil have the same refractive index.

Identifying Parts of the Compound Light Microscope

Locate and **identify** the following parts of the microscope in **Figure 4.1.**

1. **Light Switch**—Located on the base of the microscope.

2. **Light Source** (Illuminator)—Provides the light that illuminates the specimen. This is located in the base of the microscope.

3. **Iris Diaphragm**—Enables the viewer to adjust the amount of light that reaches the specimen. This is located below the condenser. A short black lever opens or closes the diaphragm.

4. **Condenser**—Focuses or condenses the light. This is located underneath the hole in the stage.

5. **Condenser Control Knob**—Adjusts the height of the condenser. Use the metal knob, located in front of the coarse and fine adjustment knobs.

6. **Mechanical Stage**—A platform with a slide holder. Serves to hold slides in place.

7. **Stage Control Knobs**—Used to move the slide around on the stage. Located hanging underneath the stage. Note: there are two knobs that control movement in two directions.

8. **Objective Lenses**—Set of three or four lenses located directly above the stage. The magnification of each lens is engraved on the metal lens housing.
   a. **Scanning Power**—This lens magnifies 4× and is the shortest of the three lenses. It is used for initial focusing and viewing.
   b. **Low Power**—This lens magnifies 10×.
   c. **High Dry Power**—This lens magnifies 40×.
   d. **Oil Immersion Lens**—This lens magnifies 100×. Use only with the help of the instructor.

9. **Revolving Nosepiece**—Allows the objective lenses to move into position above the specimen. Connects the lenses to the lower end of the body of the microscope.

10. **Body Tube**—A metal casing through which light passes to the oculars.

11. **Ocular (Eyepiece)**—This lens has a magnification of 10×, and is located in the uppermost part of the microscope. One of the oculars has a pointer that can be moved by turning the tube the ocular is housed within. If you have a binocular microscope, you can adjust the distance between the lenses to accommodate your eyes.

12. **Arm**—The upright structure attached to the base used in carrying the microscope.

13. **Base**—The heavy, flat support on which the microscope rests.

14. **Coarse Adjustment Knobs**—Used to rapidly alter the distance between the objective lenses and the stage to focus on an object. These are the large knobs that extend from each side on the lower part of the arm.

15. **Fine Adjustment Knobs**—Used to slowly alter the distance between the objective lenses and the stage when using the longer objective lenses (40× and 100×). These are the smaller pair of knobs that extend from each side of the lower arm.

Label this diagram fully.

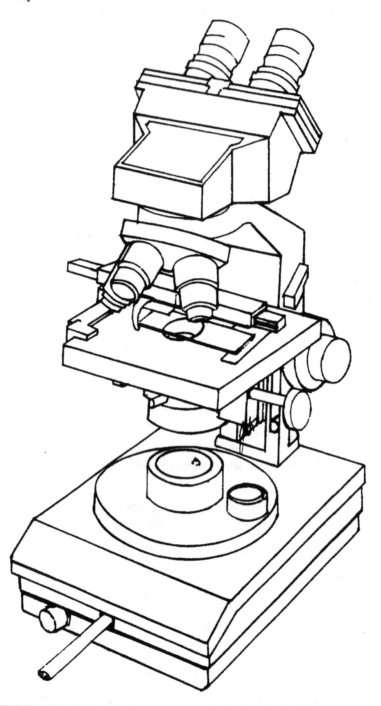

Figure 4.1: Schematic of a Binocular Microscope

Use and Care of the Microscope

1. Remove your assigned scope from the cabinet or shelf and using both hands carry it to your desk. The stage should face the user.

2. Check to see that the scope was put away properly by the previous user (see below) and report at once any scopes that were not.

3. Remove the cord carefully from the back of the stage and plug it in. Turn the light on.

4. Place the slide on the stage using the **mechanical stage.** Lower the 4× objective to its lowest point. Check to see that both oculars are equally adjusted.

5. When focusing on slides, always start with the 4× objective at its lowest point and focus by turning the coarse adjustment up away from the slide. Fine tune the image by using the fine adjustment. Use the iris diaphragm to adjust the amount of light to give the best resolution.

6. When switching to higher powered objectives, DO NOT touch the coarse adjustment. The microscope is made so that only small changes in the fine adjustment and the amount of light are necessary.

7. ALWAYS turn back to 4× before removing a slide from the scope.

8. When you are finished with the scopes for the day, complete the following steps.
   a. Turn the 4× objective in place and raise it to its highest point.
   b. Turn off the light.
   c. Moisten lens paper with lens cleaner and wipe off the objective and ocular lens with it.
   d. Then wipe off the stage and the adjustment knobs with a piece of lens paper.
   e. Polish all of these with a dry piece of lens paper.
   f. Place the cord at the back of the stage.
   g. Show the microscope to your instructor before returning it to the proper cabinet or shelf.

Using the Compound Light Microscope

1. Carry the microscope to the lab bench using **TWO HANDS. Never tilt the microscope, as this may cause the ocular lens to fall out.**

2. With the condenser control knob, move the condenser up to the hole in the stage. You should see the light shining through the condenser. Open and close the iris diaphragm using the protruding lever. Note the change in light intensity. Now open the iris diaphragm completely to allow for a maximum light.

3. Turn the objective lenses so that the scanning power lens (4×) is clicked into place.

4. Obtain a prepared slide of the letter "e." On the stage is a slide holder. Open the metal arm to make room for your slide. Gently place the slide into the holder, nestling the slide towards the back. Gently allow the metal arm to close on the slide.

5. Practice moving the slide on the stage using the stage control knobs. Manipulate the slide so that the letter "e" is positioned directly over the condenser.

6. With the coarse adjustment knob, move the 4× objective all the way down, towards the slide. Now look through the ocular, and with the coarse adjustment knob, focus on the letter "e". (This is a good time to adjust the oculars for your eyes, if you have a binocular microscope.)

7. Move the "e" so that it is in the center of the field. Adjust the light intensity, if necessary.

   a. Move the slide to the LEFT. In which direction does the letter move? _____

   b. Move the slide to the RIGHT. In which direction does the letter move? _____

   c. Move the slide AWAY from you. In which direction does the letter move? _____

   d. Compare the orientation of the letter "e" when viewing it with the naked eye, then through the microscope. _____

   _____

   e. Draw the letter "e" below.

   Letter "e": Magnification _____ ×

8. **WITHOUT TOUCHING THE COARSE FOCUSING KNOB,** turn the low power objective until it clicks into place. Using the coarse or fine adjustment knob, refocus on the letter. (Remember, the lenses are parfocal.)

   Draw the letter "e" below.

   Letter "e": Magnification _____ ×

9. Before going to the next lens, position the "e" so that part of the letter is directly in the center of your field of view. Now turn the higher power (40×) objective until it clicks into place. Using the **FINE ADJUSTMENT KNOB ONLY,** refocus on the letter.
   a. Draw the letter "e" below.

Letter "e": Magnification _____ ×

b. What has happened to the field of view? _____

c. What has happened to the letter "e"? _____

d. What has happened to the light intensity? _____

e. Why do you only use fine adjustment when viewing an object under high power?

_____

Total Magnification

The **total magnification** of the object can be calculated by multiplying the power of the ocular by the power of the objective.

Ocular	×	Objective	=	Magnification
_____	(Scanning)	_____		_____
_____	(Low)	_____		_____
_____	(High-dry)	_____		_____
_____	(Oil immersion)	_____		_____

Using Prepared Slides

Some of the slides you will be using in lab are prepared slides, like the letter "e" and colored-threads slides. Obtain the following slides, and draw a representation of each specimen in the space provided. You will have to determine the best magnification to use when drawing the object. Your instructor will help you to decide.

1. White Blood Cell: Magnification _____ ×

2. Onion root tip: Magnification _____ ×

3. Red Blood Cell: Magnification _____ ×

4. Leaf Cross Section (c.s.): Magnification _____ ×

Preparation of a Wet Mount Slide

An alternative to using a prepared slide is to make your own slide, if a biological specimen is available. The **wet mount slide** is a temporary slide and can be used on living or preserved materials.

1. Obtain a clean glass slide.

2. Place your specimen (a small amount is generally the rule) onto the center of the **slide.** (see Figure 4.2)

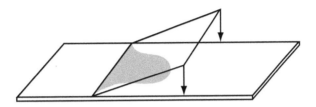

Figure 4.2

3. Place the edge of a clean **coverslip** at the edge of the specimen. Slowly lower the coverslip over the specimen. This way, no air bubbles will get trapped.

4. When you are done viewing and drawing the specimen, discard the slide as instructed.

5. What is the purpose of using a coverslip?

_____

_____

_____

6. Make a wet mount slide and draw each of the following:

   a. *Elodea* leaf: Magnification _____ ×

   b. Pond Water: Magnification _____ ×

5

Bacteria, Protista, Algae, and Fungi

Bacteria

Bacteria are single-celled organisms without a nucleus and other internal organelles. The bacterial chromosome is a circular loop of DNA, which lacks the histone-protein packaging found in all eukaryotes. Bacteria may possess additional, circular loops of DNA that are smaller than the main bacterial chromosome. These "plasmids" often carry genes that confer resistance to antibiotics. Although structurally simple, bacteria exhibit a complex metabolism. Bacteria and the photosynthetic blue-green algae are called prokaryotes. Thus, these organisms are characterized by (1) lack of a nuclear membrane (prokaryotic) and other organelles, (2) small size, (3) presence of a cell wall, and (4) a "naked" chromosome consisting of a single loop of DNA.

There is tremendous variety within bacteria. We find modes of life among them more varied than most other forms of life. There are unicellular and colony-like forms. Some are aquatic, some terrestrial. A few are autotrophic, but most are heterotrophic. Most of the heterotrophic bacteria are saprophytes.

We have time to study only the most common bacteria that also happen to be among those with the simplest forms.

Characteristics of the Most Common Bacteria

1. Unicellular and microscopic

2. Nonphotosynthetic

3. Saprophytic

4. Rigid cell walls

5. Reproduction by simple fission

Microscopic Appearance of Bacteria

1. COCCUS—A spherical-shaped cell (plural = cocci)

2. BACILLUS—A rod-shaped cell (plural = bacilli)

3. SPIRILLUM—A twisted rod-shaped cell (plural = spirilli)

Bacteria are often found in groups. When bacteria show a characteristic pattern of grouping, they are often named according to the grouping pattern.

Common groups are:

DIPLO—2 together

STREPTO—A chain of bacteria

STAPHYLO—A cluster of cells

Exercise A: Appearance of Bacteria

Examine a prepared slide or photos showing the three cell shapes found in bacteria. Draw on the answer sheet an example of each cell type when viewed under high power.

Exercise B: Bacteria on Plates

Examine plates of bacteria and compare them with yeast. Describe representative samples in your answer sheet (e.g., E. coli, B saprophytic, Serptia sp., Micro bacillus, Pseudomonas sp.).

Exercise C: Environmental Sampling

Bacteria and fungi occur throughout the biosphere. To demonstrate their ubiquity, you will expose two sterile agar plates to the environment. After a 48 hour incubation, you will examine, describe and draw any bacterial and/or fungal growth on your exposed agar plates. Do not open the plates.

Inoculating a sterile petri plate using a sterile loop or a moistened swab, sample any surface or water in the lab. While still holding the inoculating loop, open the lid of your sterile, agar plate just enough to permit access of the loop. Gently streak the surface of the agar back-and-forth over the top-third of the surface. Remove the loop and immediately close the lid, as demonstrated by the instructor. Successful streaking should produce distinct bacterial colonies that result from the repeated cell divisions of a single bacterium.

Kingdom Protista

The kingdom Protista include the simplest organisms that are eukaryotic, i.e., cells with a nucleus and other membrane-bound organelles. Protista are mostly unicellular, some are colonial, and a few are multicellular. Protists provide for their needs using all feeding strategies; some are photosynthetic, others heterotrophic, and saprophytic. Some cells are mobile, either with cilia, flagella, or amoeboid-like movements. Most biologists consider modern day protists to be the descendents of the first eukaryotic organisms on Earth. These earliest eukaryotes are undoubtedly the ancestors of all the multicellular kingdoms (fungi, plants, and animals).

To examine protists, prepare wet mounts of the following specimens, and use the scanning and low power objectives on the compound microscopes. A drop of methyl cellulose or other viscous material can be used to slow their swimming speed.

From a colony of amoeba, look for a small speck near the bottom. Make a wet mount and draw what you see. Observe how the amoeba forms a pseudopod to crawl.

Observe and draw a *Paramecium* ciliate from a colony in the lab. *Paramecium* propel themselves using the rowing motion of thousands of cilia that cover the entire cell.

Volvox is colonial. The cells live and work together, but they still retain their independence from each other. *Volvox* is also photosynthetic. Observe and draw.

Observe and draw a Euglena flagellate from the colony. These move by means of a single flagellum.

Algae—Some biologists include the algae in the Protista, others consider them to belong to the kingdom of plants. Similar to many protista, the algae are aquatic, reproduce using motile, swimming reproductive cells, and often the dominant life stage is haploid, i.e., has only one set (n) of chromosomes. Similar to plants, algae are photosynthetic. Some like the green algae contain similar pigments, whereas others such as the brown algae, red algae, and the diatoms contain other pigments that assist in light harvesting. Algae, unlike modern plants, lack vascular tissues, cell walls, and do not develop multicellular reproductive structures.

Kelp, or seaweed, or mostly members of the brown algae (Phaeophyta). Examine and draw any specimens available.

Kingdom Fungi

The fungi are multicellular, heterotrophic organisms that primarily exhibit a saprophytic life style. As such, they are some of the more important decomposers in most ecosystems. The body of a fungus consists of tubular filaments, called hyphae, that grow over and into their food. Fungi, build a cell wall to the exterior of the plasma membrane as in plants. However, the composition of fungal cell walls is more similar to the exoskeleton of insects (i.e., it consists of chitin) than to the cellulose cell walls in plants. They secrete digestive enzymes, allow the digestion to proceed extracellularly, and then absorb the digested soup through their cell walls and plasma membranes. A mat of hyphae is called a mycelium.

Fungi are classified into different classes on the basis of the reproductive structures. These include the Zygomycetes, including some species of bread mold, the Ascomycetes, or the sac fungi, such as yeast and ergot of rye, the Basidiomycetes, or club fungi/mushrooms, and a class for which reproductive structures are generally unknown, the fungi Imperfecti.

Examine the prepared slides of bread mold, a zygomycete, of an ascomycete, and a basidiomycete. Prepare and label drawings.

Lichens—Lichens are symbiotic organisms that consist of fungal hyphae that surround and trap unicellular green and blue-green algal cells. These two organisms are completely dependent upon each other to survive. The fungus provides a home, moisture, and inorganic nutrients for the algae, whereas the algae produce sugars via photosynthesis for the fungus. Lichens are important colonizers of bare rock. They contribute to soil formation by secreting extracellular acids that degrade rock. Lichens are also very sensitive to air pollution, and are used as air quality bioindicators. In other words, if lichens are present, then the local air is relatively pollutant free.

Examine the three types of lichens and fungi, and draw their distinct body types.

EXERCISE 5

Data/Answer Sheet 1

Bacteria

A. Appearance of Bacteria slides

Drawings from micrographs or slides

    coccus   *bacillus*   *spirillium*

B. Bacteria on Plates or tubes — Draw and/or describe

C. Environmental Sampling of Bacteria

Draw the appearance of your agar plates following exposure and incubation

Exposure: _____

EXERCISE 5

Data/Answer Sheet 2

Protista

Draw and label:

Amoeba   *Paramecium*

Volvox

Algae   *Euglena*

Data/Answer Sheet 3

Fungi

Draw and label:

Bread mold *Rhizopus* (Zygomycete)

Ascomycete

Basidiomycete

Lichens

crustose   foliose   fruiticose

6

Plant Diversity

All plants are multi-celled (cells are differentiated in tissues), and most are photosynthetic eukaryotes that are capable of producing food from water, carbon dioxide, and dissolved minerals, using sunlight as the energy source. Photosynthetic bacteria and protistans (algae), and green plants are the primary producers for nearly all communities. In addition to the ability to trap and use sunlight to synthesize glucose, the photosynthetic organisms are the source of atmospheric oxygen. The multicellular Kingdom Plantae colonized terrestrial habitats. Major evolutionary trends within the plant kingdom include: the development of vascular tissues, the increasing importance of the sporophyte generation, the development of seeds, and lastly the emergence of flowering plants. Most of these trends are characterized by the development of traits that enable plants to live and reproduce under drier and drier conditions.

Kingdom Plantae

    I. Bryophyta - Lack vascular tissue
       A. Mosses
       B. Liverworts
    II. Tracheophyta - Possess vascular tissue: Xylem and Phloem
       A. Horsetails and Ferns
       B. Gymnosperms - Cone bearing plants
       C. Angiosperms - Flowering plants
          1. Monocotyledons (1 seed leaf)
          2. Dicotyledons (2 seed leaves)

A. Nonvascular Land Plants (Bryophyta): Liverworts and Mosses

Liverworts. These plants have invaded the land but still need a film of water for survival and sexual reproduction. This is necessary so that the **motile sperm** can swim to the egg-bearing

From *Concepts of Biology Laboratory Manual*, 3rd Edition, by Boise State University. Copyright © 2002 by Boise State University. Reprinted by permission.

archegonium of the female specimen. The **male plant** has heads which are shallowly indented whereas the female plant has stalked heads which are deeply cleft.

The body of the liverwort is called a **thallus** and is multilobed.

Male   Female

Mosses. These bryophytes are often found under drier conditions than liverworts, but sometimes certain species grow together with liverworts. The gametophyte generation is the dominant generation and is what we are observing when we seen green moss growing on a moist rock. Mosses also grow as separate male and female plants. It is only the female that produces the stalked sporophyte with the capsule containing the spores which result from sexual reproduction. The male plant produces sperm which swim to the female gametophyte and fertilize the egg. The zygote then grows into a sporophyte with a terminal capsule. Certain sporophyte cells within the capsule undergo meiosis to produce $1n$ spores which, when released, can grow into new moss plants (½ male, ½ female).

Exercise: Draw and label the parts of the moss provided.

**Sporophyte
(2n)**

**Gametophyte
(1n)**

B. Vascular Seedless Plants: Ferns and Fern Allies

A major evolutionary advance in the design of green plants occurred with the appearance of vascular tissue. Vascular tissue allows for the transport of water and solutes dissolved in the water over great distances within the plant body. The presence of the vascular tissues allowed for a dramatic increase in plant size and decreased the dependence on water. However, a film of water is still required for sexual reproduction in the seedless vascular plants. A second major change is the appearance of an independent sporophyte generation, with the gametophyte being reduced to a physically insignificant body (see below).

Whisk Fern. *Psilotum* is a living fossil since few of its kind are alive today, but in earlier times they were one of the most common plants. This is a tropical plant which occurs in Florida. It is one of the earliest plants to possess **xylem** and **phloem,** vascular tissues involved in support and translocation of water and nutrients. These plants still require a film of water for sexual reproduction because they too possess motile sperm.

Horsetails. These forms are quite common in Idaho. They are also commonly called snakegrass and scouring rushes. The stems contain silica which makes them rather abrasive. Early pioneers used them to clean pots (a recyclable SOS pad) and accordingly referred to them as scouring rushes.

Ferns. The ferns are a prominent group of plants because of their size and the number of species. The undersides of the compound leaves typically produce the rust-brown spores which germinate to produce the fern gametophyte. The leaves are called *fronds*. They possess sori (containing the spores) on the underside of the fronds.

Fern Prothallus. Fern gametophytes bear structures that produce egg and sperm cells. Ferns also require films of water for sexual reproduction. They produce motile sperm which swim to fertilize the egg.

Exercise: Draw and label the parts of the fern provided.

C. Seed Plants: Gymnosperms ("naked" seeds) and Angiosperms

The next major event in the evolution of plants is the appearance of seeds. The presence of the seed totally eliminates the direct role of water in plant reproduction. Seed-bearing plants (gymnosperms and angiosperms) are the most successful of all the land plants. You can easily verify this by simply looking at the plants that are all around you. Most of these plants are either **gymnosperms** ("naked" seeds) such as conifers like pines, spruces, firs, and junipers or **angiosperms** ("protected" seeds), the flowering plants. Seeds arise from mature, fertilized **ovules.** Ovules contain egg cells and are surrounded by a protective tissue called the **integument.**

Following fertilization, the ovule will develop into a seed. Within the seed an **embryo** begins to develop and is protected from harsh environmental conditions as the integument develops into the **seed coat.** In pines, seeds are borne on the scales of the female or ovulate cones. Both female and male cones are provided for your inspection. The male cones are not permanent structures. They produce the yellow, powdery pollen that often covers cars, tents, and water surfaces during the spring in this area.

Ovulate and Staminate Cones. Pines, spruces, firs, junipers, etc. typically produce female (ovulate) and male (staminate) cones. The staminate cone releases the pollen which is windborne to the female cone. Some species have both male and female cones on one plant (**monoecious**) whereas others have them on separate plants (**dioecious**). The yellow, powdery pollen is often seen covering cars, tents, and water surfaces during the spring in this area.

Seed and Embryo. The seed is a specialized unit for the protection of the embryo. The pine seeds are borne within the female or ovulate cones. The hard coat is formed by the parent sporophyte. The nutritive tissue is produced by the female gametophyte within which is embedded the embryo of the next sporophyte generation.

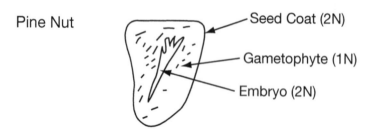

Pine Nut

Seed Coat (2N)

Gametophyte (1N)

Embryo (2N)

Conifers. Six different common types of gymnosperms are commonly planted as ornamentals. Some do not have the cones which typify most conifers (cone-bearing gymnosperms).

1. **Pines** usually have two or more needles borne in a cluster. Ponderosa pine typically has three, lodgepole pine two, and white pine five.

2. **Spruces** have needles which are stout (almost square in cross section), sharp tipped and single.

3. **Firs** have needles that are flat, round tipped and not so rigid. They are also borne singly on the stem.

4. **Junipers** don't have needles. The scales are of two types with most generally rounded and overlapping to form green, scaly branches. Some scales, however, are sharp so you do not usually grab juniper branches.

5. **Arbor Vitae.** This very common ornamental is often called a cedar, but it is not a true cedar. The branches grow in a flattened manner and readily fit in an overlapping pattern.

6. **English Yew.** A locally widely-planted ornamental. The sexes are separate. The red arils (fleshy seed cup) are found only on the female plants. The foliage and berries are toxic to humans.

Exercise: Draw and label the parts of the conifers provided.

Seed Plants: Angiosperms (flowering plants, "protected" seeds)

The last significant evolutionary event was the appearance of flowers. Flowers are usually the first thing we think of when we hear the word plant. The parts of the flower are highly modified leaves. **Angiosperms,** or **flowering plants,** are the most successful terrestrial plants on earth. They have dominated the land for more than 100 million years and they occupy a wide variety of habitats. They live in very arid locations, in wetlands, freshwater, and seawater. Plants also possess a great deal of morphological diversity. The flower is the reproductive structure of the plant. In this exercise you will be introduced to terms that describe the various parts of the typical flower. See Figure 6.1 for the generalized structure of a dicot flower. Examine the flowers provided by the laboratory instructor and locate the floral structures indicated below.

Flower Structure. Study the diagram (p. 160 and the flower model) and the representative flowers so that you can identify the major parts: **receptacles, sepals, petals, ovary, stigma, style, carpel, anther, filament, stamens.**

Fruits. A fruit is the expanded and **ripened ovary** of one or more carpels. The fruit serves to protect the seeds and promotes seed dispersal from the parent plant. Fruits can be classified as either **simple, aggregate,** or **multiple** fruits. Also, fruits can be either **fleshy** or **dry.** Observe the fruit types assembled at this station.

Stem. The cross *section of the stem* shows an outer epidermis with a thin layer of **cortex.** There is then a thin layer of **phloem** followed by a thin layer of **xylem.** The larger cells in the center comprise the **pith.** The plant is alfalfa.

Leaf. The microscope shows a cross section through a representative leaf of a flowering plant. *Using the chart identify the following structures:* **epidermis, stomata, vascular tissue, xylem, phloem, palisade mesophyll, spongy mesophyll.**

Exercise: Draw and label the angiosperm specimens provided, using the diagram & flower models provided

Figure 6.1: Generalized Structure of a Dicot Flower

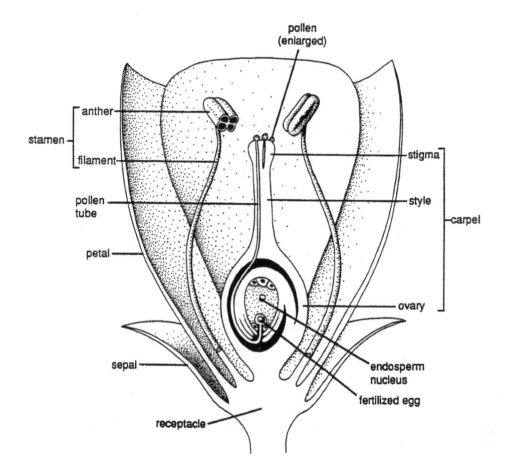

7

Natural Dyes Research (Inquiry-Based)

Introduction

Throughout the world there is a wide diversity of plant species which are now being destroyed by deforestation at a phenomenal rate, especially since we still have not been able to effectively combine development of plant resources and conservation. Of the many plant species, which have been identified to date, only some have been fully exploited for their biological and chemical potential as medicines, pesticides and natural dyes etc.

Natural dyes have been used by all peoples of the world. The Egyptians, Chinese and Indians have been using natural dyes for millenia, and it has been established as a craft in India, China and South America since. The earliest record of the use of dyestuff comes from China sometime before 3000 B.C. It it known that Egyptian mummies some 3000 years ago have been wrapped in a blue colored linen. In America, it is reported that Native Indians used various colors even before the advent of the Europeans. Perkins produced the first synthetic dye in 1856, and relatively cheap, commonly available dyes became available in the late 19th and 20th century. Now because of the health concerns of these synthetic dyes, many dyers are going back to the use of natural dyes.

A large number of plants can be used to extract vegetable dyes, for example onion skin, beetroot, carrot, rhubarb, spinach, colorful flowers and berries such as cranberries, blackberries and blueberries; tea, coffee, red cabbage, and turmeric. In this research you are required to use the inquiry-based methodology to investigate several plant materials for their dyeing capability with and without mordants, on different fabric. Mordants are chemicals that aid the chemical reaction between the dye and the fabric and can also alter and enhance the color, e.g., alum, iron, cream of tartar and urine.

Task: You are given the job of investigating a number of natural dyes to find out if they are suitable with particular fabric.

First, conduct library research into natural dyes. Next, have a class discussion facilitated by the instructor on biodiversity utilization and conservation, especially natural dyes.

Working in groups of three or four, use the following information to design your experiments to extract and test the dyes. You will be observed and graded on your individual involvement and on your interaction in your group.

Preparation of the Dye

Materials:

1. Spatula, beakers, water, alum, cream of tartar, hot plate, spoons, paper towels, baking soda, vinegar, and droppers.
2. pipettes.
3. measuring cylinders.
4. scissors.
5. funnels.
6. cheesecloth.
7. balance.
8. weighing boats.
9. glass rods.
10. pH papers.
11. thermometer.
12. gloves.
13. goggles.

Method:

1. Cut the plant, soak, boil in water for 15 minutes, (e.g., 5 g plant material/50 ml water) and filter.
2. The filtered liquid is the dye bath.
3. Record the color obtained.

Dyeing the Fabric

Materials:

1. Alum.
2. iron (ferrous sulfate).
3. cream of tartar.
4. white cotton cloth, wool.
5. linen.

6. silk fabric.

7. lemon; ammonia.

8. string.

9. pegs.

Method:

Dye pieces of the material (e.g., 8 cm × 8 cm) by boiling in the dye bath (e.g., 10–20 ml) for 10 minutes with or without different mordants e.g., alum, iron, and cream of tartar; dry the fabric in the lab, and record all observations including the color of the fabric obtained. Rinse dyed material in warm water until water runs clean. Wash with mild detergent and rinse again, then dry on a line.

You can investigate the effect of cold and hot extraction; effect of boiling time; effect of different fabric e.g., cotton, wool, linen, silk; effect of acid (lemon—few drops) and alkali (ammonia—few drops); and effect of different types of mordants or different quantities of mordants (e.g., 0.1 g /ml of dye) and if time permits the effect of washing (wash fastness); and rubbing (rub fastness).

Wash fastness can be tested by placing a dyed material after rinsing and washing, in a jar with mild soap and warm water and then shake, to see whether the color will bleed into the water. Alternately, place wet cloth on a white cloth and observe any bleeding of the color.

Rub fastness can be tested by rubbing the dyed material with a clean piece of white material. Any non-fast dye is picked up by the white material.

Laboratory safety: remember to use goggles, and observe all safety procedures relating to the use of heat, and chemicals e.g., avoid getting any solutions on your skin, and use a pair of gloves.

Report—Write up a full experimental report about the suitability of each dye using the format as instructed before, i.e., title; introduction; materials, methods; results and observations; discussions; conclusion; bibliography. Your report, which should be typed double-spaced, will be graded using a rubric assessment tool (Figure 20.1).

Bibliography

1. Dean, J. (1999) Wild Color. Watson-Guptill Publications. New York.

2. MacFoy, C. & V. Pratt. (2003) Laboratory Investigations of Selected Natural Dye-Yielding Plant Species. Colourage, vol 50, AN.REP.

Name: _____   Date: _____

Lab Section: _____

Lab Activity: _____

Date: _____

Name(s)	Never 0	Rarely 1	Sometimes 2	Usually 3	Always 4	Score Weight	Total Score
A. Connection							
• Demonstrates scientific inquiry						2	
• Correctly incorporates appropriate background research						3	
• Shows relationships between background research and the experiment						3	
• Employs sufficient background research						2	
• Correctly references data/research						2	
• Uses scientific vocabulary correctly						2	
B. Process							
• Correctly follows the procedure						3	
• Collaborates with team						2	
C. Cognition							
• Appropriately and correctly displays data in the form of a chart, graph or table						3	
• Correctly analyzes and justifies the results						4	
• Forms reflective conclusion that links new and prior knowledge						4	
• Presents a comprehensive analysis and summary of the data.						3	

FINAL SCORE = _____ POINTS

Instructor Comments

Survey of the Animal Kingdom

8

Introduction

Members of the animal kingdom are multicellular, eukaryotic organisms that lack cell walls. Animals are *heterotrophs*. Also, most animals are motile with nerves and muscles, and they occur in water, on and under land, and in the air.

The animal kingdom is considered to be *monophyletic*, i.e., all animals are descended from one common ancestor. This ancestor was most likely a single-celled, flagellated, motile protistan, that adopted a colonial lifestyle and eventually a multicellular lifestyle. In the animal body, hundreds to trillions of cells work interdependently with all other cells to achieve homeostasis, grow, and reproduce.

The morphological innovations that occurred throughout animal evolution are evident in modern-day descendents of some of these early animal forms. The *phylum Porifera*, sponges, are multicellular, yet cells are not arranged into tissues. Also, sponges lack symmetry.

The *phylum Cnidaria* are some of the simplest and earliest in terms of tissue construction and symmetry. Cnidarians like jellyfish and corals, are *radially symmetrical*. They lack a front (head) and a rear (tail). Muscle tissues are not developed; hence, these animals consist of only two tissues layers *(ectoderm and endoderm)*. Cniarians are thus *diploblastic* (two-layered). Radial symmetry is suitable for animals that are sessile, or drift.

A *bilateral symmetry* seems better suited for mobility. Bilateral animals usually have a head region *(cephalization)* with the sense organs and mouthparts on the front-end of the body that first encounters the world as the animal moves through it. Bilateral animals are composed of three tissue layers (the third layer being *mesoderm* which develops into muscle). One of the simplest bilateral animals are the flatworms, or *phylum Platyhelminthes*. Flatworms have a head region with large eye spots, yet their organ systems are poorly developed. Similar to Cnidarians, flatworms have a digestive sac with only one opening serving both as mouth and anus. Also, the interior of the body is solidly filled with tissue, i.e., a condition called *acoelomate*.

"Survey of the Animal Kingdom" from *Introductory Biology Lab Manual* by Jim Goodwin, Mark Barnby and Carol Dixon. Copyright © 2000 by James Goodwin, Mark Barnby and Carol Dixon. Reprinted by permission of the authors.

With the development of internal organ systems, a fluid filled cavity evolved to serve as a cushioned compartment for organs of digestion, excretion, and reproduction. The *phylum Nematoda,* roundworms, and the *phylum Rotifera,* are some of the earliest animal groups to have a *complete digestive tract* with a separate mouth and anus. Also, a fluid filled coelom has developed, however, this cavity is rather primitive and lacks a lining and tissue supports for organs. Such cavities are known as *pseudocoeloms,* i.e., false cavities.

More advanced animal phyla have a body cavity that is lined and supports organs with tissue. These animals are known as the *coelomates,* or *eucoelomates,* i.e, those with a true coelom. Depending on the pattern of embryonic development, most higher invertebrate phyla belong to the *protostomates* (protostome = first mouth), whereas the vertebrates and related phyla belong to the *deuterostomates* (deuterostome = second mouth). Protostomates include the phylum Mollusca (chitons, snails, bivalves, and cephalopods), and the segmented phyla, the *Annelida* and the *Arthropoda.* Annelids, the segmented worms, include the free-living terrestrial earthworms, their marine cousins the polychaetes, and the ectoparasitic leeches. Annelids rely on a hydrostatic skeleton for support and a tough outer cuticle. They also have closed circulatory systems. The phylum Arthropoda is the most successful animal phylum in the history of the Earth (see Table below). Arthropods have an exoskeleton that must be shed to permit growth. Arthropods have jointed appendages that may be modified to perform various functions such as feeding or sensory detection. Some arthropods (mostly insects) undergo a radical re-design in their body form during their development, i.e., metamorphosis.

Deuterostomates include the invertebrate *phylum Echinoderms* (starfish, sea urchins, and sea lilies). These marine, mostly sessile creatures with their radial symmetry appear to be primitive, but the bilateral design of the larval stage (the stage that disperse the animal to new habitats) and the deuterostome embryonic development suggest that this phylum is rather advanced evolutionarily. The *phylum Chordata* (which contains the *sub-phylum Vertebrata)* include animals with a dorsal support rod, notochord, dorsal, hollow nerve cord, a post-anal tail, and segmented muscles (somites). Vertebrates replace (surround) the notochord with a bony vertebral column.

Kingdom Animalia

Invertebrates	Vertebrates
Phylum Porifera—5,000 species	Subphylum Vertebrata—>44,000 species
Phylum Cnidaria—9,000 species	Class Agnatha—63 species
Phylum Platyhelminthes—13,000 species	Class Chondrichthyes—850 species
Phylum Nematoda—12,000 species	Class Osteichthyes—20,000 species
Phylum Rotifera—2,000 species	Class Amphibia—3,900 species
Phylum Mollusca—110,000 species	Class Reptilia—6,000 species
Phylum Annelida—12,000 species	Class Aves—9,000 species
Phylum Arthropoda—>6 million species	Class Mammalia—4,500 species
Phylum Echinodermata—6,550 species	
Phylum Hemichordata—90 species	
Phylum Chordata	
Subphylum Urochordata—1,250 species	
Subphylum Cephalochordata—23 species	

Exercise A: Comparative Survey of Animal Phyla

When comparing example organisms from some of the major phyla of animals, pay attention to the following characteristics, and summarize their status in the table provided on the Data/ Answer Sheet.

- *Tissue Organization:* Are cells organized into tissues, and if so how many tissue layers are found (diploblastic vs. triploblastic)?

- *Symmetry:* Does the animal lack symmetry, or is it radially or bilaterally symmetrical?

- *Digestive Openings:* Is there one external opening that functions both as mouth and anus, or are there two separate openings?

- *Body Cavity:* Is there a fluid-filled space inside the outer body wall that is either lined or unlined with tissue (eucoelomate or pseudocoelomate, respectively)? Or, is the interior of the body solidly filled with cells (acoelomate)?

- *Segmentation:* Does the body design of the animal represent the linear repeat of similar body parts or segments?

- *Support System:* How does the animal support itself against gravity, and provide attachments for muscles to allow for posture and locomotion? Does the animal have hydroskeleton, an exoskeleton, or an endoskeleton?

- *Appendages:* Are appendages present (attachments to the body trunk)? Do they occur all along the body, or are they restricted to certain body regions? Are all the appendages similar, or are some modified for a different use?

- *Locomotion:* Does the animal walk, swim, drift, or fly?

- *Habitat:* Is the animal terrestrial, subterranean, or spend most of its time suspended in the atmosphere? Is the animal aquatic, either marine or freshwater?

Procedure to Viewing the Animal Specimens

1. Phylum Porifera—Use a compound light microscope to examine prepared slides of longitudinal sections from the sponge *Grantia*. You should see choanocytes (collar cells), amoeboid sex cells, and spicules. In a longitudinal section, or in preserved whole sponges, you should see the opening or osculum.

2. Phylum Cnidaria—Use an eyedropper to transfer living *Hydra* onto a watch glass or a depression slide. View *Hydra* using a compound dissecting microscope. Add brine shrimp or water fleas and watch the *Hydra* feed using its tentacles. Prepare a wet mount of

Hydra on a slide using a coverslip. Under a compound microscope, examine the tentacle and observe the swollen stinging cells or cnidocyte. Use a compound microscope to examine prepared slides of cross sections (x.s.) of *Hydra*. Note the presence of two tissue layers, i.e., diploblastic. Examine pieces of coral skeleton. What is this made of?

3. Phylum Platyhelminthes—Use an eyedropper to transfer living *Dugesia,* a turbellarian, or free-living, flatworm to a watch glass or depression slide. If you shade half of the dish with a piece of paper, how does *Dugesia* respond? View a prepared slide of a x.s. of *Dugesia.* Note the presence of a third layer of tissue, muscle (triploblastic). Also, note the acoelomate body construction of *Dugesia.* Examine the prepared specimens of tapeworms.

4. Phylum Nematoda—Examine a prepared slide of x.s. of the roundworm *Ascaris* using a compound microscope. Note its triploblastic tissue arrangement, and the digestive tract, and its fluid-filled, but not tissue lined body cavity (pseudocoelom). On whole, preserved specimens, note the separate mouth and anus.

5. Phylum Mollusca—Examine the preserved specimens of snails, bivalves, and cephalopods. Note that the non-segmented body is covered by an exoskeleton (the shell).

6. Phylum Annelida—Use the compound microscope to examine prepared slides of x.s. of *Lumbricus,* an earthworm. Note the circular and longitudinal muscles, the digestive tract, and the fluid-filled coelom (eucoelom) lined with tissue. Examine whole, preserved specimens of polychaetes. Transfer a living leech into a watch glass and observe its large sucker used to attach itself to a host to suck blood.

7. Phylum Arthropoda—Use a dissecting microscope to examine the external body parts of a crayfish (Class Crustacea). Locate the following structures: antennae, antennule, eyes, carapace, cheliped (walking leg 1), walking legs 2 → 5, abdominal segments, uropod, and telson. Similarly, examine a grasshopper (Class Insecta) and locate the following structures: antennae, ocellus, compound eyes, mandible, labrum, feeding palps, prothorax, mesothorax, metathorax, wings, legs, spiracles, tympanic membrane, abdomen, and ovipositer.

8. Phylum Echinodermata—Examine a preserved starfish. Locate the dorsal anus and the ventral mouth. Along the ventral side of the five arms are tube feet.

9. Phylum Cordata/Subphylum Vertebrata—Examine a skeleton of a small mammal. How many bones do you recognize? Are they homologous to bones of the human skeleton? The diets of mammals can be deduced from an examination of their tooth structure. Unlike reptiles where all teeth are similarly shaped (homodont), mammal tooth design varies depending on the diet (heterodont). Observe the skulls and jaws of various mammals. Can you determine which are herbivores, carnivores, or omnivores?

Exercise B: Key Innovations in Animal Evolution

First identify and label the organisms on the margins of the diagram. Using the summary information of animal characteristics from the introduction to this chapter, fill in the blank boxes on the family tree of animals with the key innovations in animal evolution.

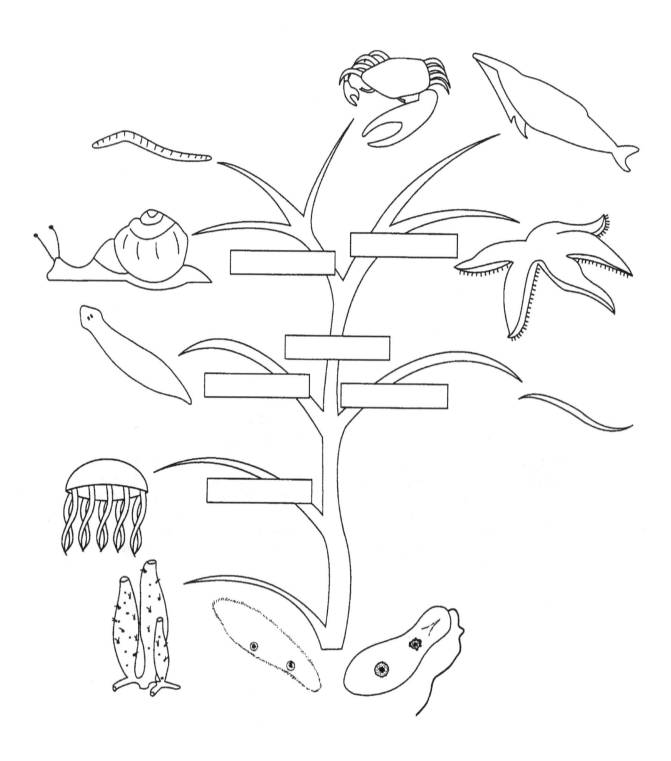

Name: _____   Date: _____

Data/Answer Sheet 1

Comparative Survey of Animal Phyla

Summary Table of Animal Characteristics

Phylum	Tissue Organization	Symmetry	Digestive Openings	Body Cavity	Segmentation	Support System	Appendages
Porifera							
Cnidaria							
Platyhelminthes							
Nematoda							
Mollusca							
Annelida							
Arthropoda							
Echinodermata							
Chordata							

EXERCISE 8

Data/Answer Sheet 2

Summary Table of Animal Habits

Phylum	Type of Locomotion	Habitat
Porifera		
Cnidaria		
Platyhelminthes		
Nematoda		
Mollusca		
Annelida		
Arthropoda		
Echinodermata		
Chordata		

9

Biological Molecules

Introduction

All matter is composed of chemically pure substances known as **elements.** Elements, in turn, are composed of **atoms.** Atoms of different elements differ from one another in the numbers of protons, neutrons, and electrons they contain. In the physical world there are **92** naturally occurring elements. Not all of these elements, however, are found in living things. Instead, the bodies of living things are composed primarily of four elements: oxygen (**O**), carbon (**C**), hydrogen (**H**), and nitrogen (**N**). Atoms of these elements and several others are chemically combined with one another in an enormous number of different combinations to form the **molecules** upon which life is based. Molecules that are composed of two or more different kinds of elements are referred to as **compounds.**

Compounds are generally divided into two major groups: **inorganic** and **organic** compounds. **Inorganic** molecules are small, simple molecules that never contain carbon atoms arranged in chains or rings. Water (**H_2O**), the most abundant compound in living organisms, is an inorganic compound. Organisms are approximately 70-90% water by weight. **Organic** compounds are large, complex molecules that always contain carbon atoms, often arranged in chains or rings. If the water is ignored, the bulk of the dry weight of living organisms is made of organic compounds.

Although the general shape of an organic molecule is determined by its arrangement of carbon atoms, or **carbon skeleton,** its chemical properties are determined by groups of atoms that are attached to the carbon skeleton. These groups of atoms are referred to as **functional groups.** Organic molecules can be categorized based on the types of functional groups that they contain. Some of the more common functional groups are shown below.

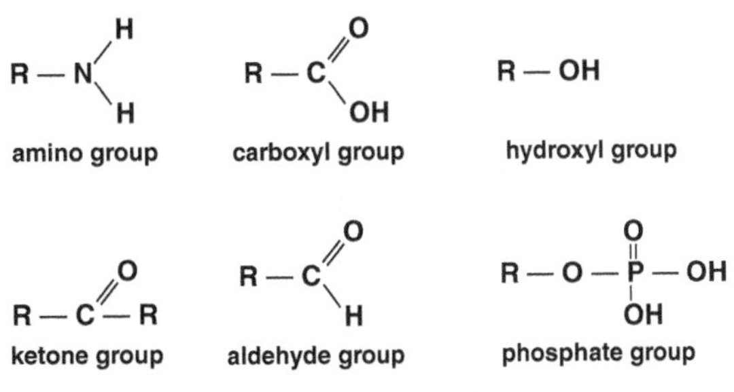

Figure 9.1: Some Common Functional Groups

From *A Look at Life: Exploring the Unity of Organisms,* 4th Edition by Crowder, Durant, and Penrod. Copyright © 2000 by Carol S. Crowder, Mary A. Durant, and Shelley W. Penrod. Reprinted by permission of the authors.

As mentioned above, the types of functional groups present in a molecule determine its chemical properties. For example, carboxyl and phosphate groups tend to make a molecule slightly **acidic.** On the other hand, amino groups tend to make a molecule slightly **basic.** Other functional groups such as hydroxyl groups, ketones, and aldehydes tend to be extremely **polar.** Molecules that contain these functional groups tend to dissolve easily in water, a polar solvent, and are referred to as **hydrophilic** ("water-loving"). Sugars are an excellent example of hydrophilic molecules. Other organic compounds consist primarily of carbon and hydrogen with few other functional groups present. They are referred to as **hydrocarbons.** These molecules are extremely **nonpolar** and generally do not mix well with water. They are referred to as **hydrophobic** ("water-fearing"). Fats and oils are good examples of hydrophobic molecules.

There are four major classes of organic macromolecules found in living things. These include **carbohydrates, lipids, proteins,** and **nucleic acids.** In today's lab, some of the properties of the first three groups will be studied. Nucleic acids, composed of monomers called nucleotides, will be studied in more detail in later labs.

Carbohydrates

There are three different classes of carbohydrates. The simplest carbohydrates are sugars known as **monosaccharides.** These "simple sugars" are the primary fuels that cell break down for energy. In addition, they serve as monomers to build larger, more complex carbohydrates. Examples of monosaccharides include glucose and fructose. **Disaccharides,** or "double-sugars," are made of two monosaccharides joined together. Sucrose (common table sugar) is a common example of a disaccharide. Figure 9.2 illustrates the formation of sucrose from the monosaccharides glucose and fructose by dehydration synthesis.

GLUCOSE   FRUCTOSE   SUCROSE

Figure 9.2:   Formation of Sucrose by Dehydration Synthesis

Polysaccharides, or "complex carbohydrates," are polymers composed of long, coiled chains of monosaccharides. Figure 9.3 illustrates a molecule of starch, a common polysaccharide used by plants to store food. (In this figure, each hexagonal ring represents a glucose unit.) Another polysaccharide, cellulose, is the main structural material found in plant cell walls. Wood, paper, and plant fibers (like cotton) are all composed primarily of cellulose.

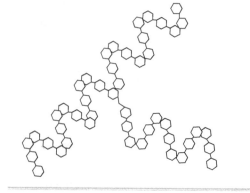

Figure 9.3:   Starch

Lipids

Lipids are a very diverse group of organic molecules. The one distinguishing feature that all lipids share in common is that they are completely or mostly hydrophobic. This is because they consist almost entirely of carbon and hydrogen with no polar functional groups attached. Consequently, lipids are extremely nonpolar molecules and will not dissolve in a polar solvent like water.

The most common lipids are known as **triglycerides.** They are macromolecules composed of a molecule called **glycerol** joined to three molecules called **fatty acids.** Figure 9.4 illustrates a triglyceride molecule. Triglycerides that are a solid at room temperature are known as **fats,** while those that are liquid at room temperature are known as **oils.** The primary function of triglycerides is energy storage. Triglycerides can store more energy than carbohydrates because they contain a larger number of energy rich carbon-hydrogen bonds. In addition, because triglycerides are hydrophobic, they can be stored with little or no water present to add additional bulk.

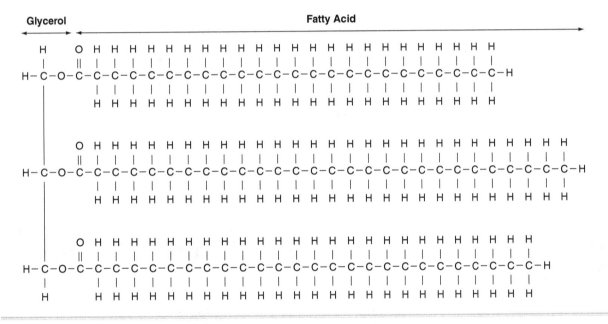

Figure 9.4: A Triglyceride Molecule

Proteins

Proteins are the most diverse of all biological molecules. Proteins are polymers composed of monomers called **amino acids.** There are **20** different amino acids found in the proteins of living things. These amino acids can be linked together in an almost infinite variety of different combinations to form all the different types of proteins found in living organisms.

All twenty amino acids have a central carbon atom with a hydrogen atom, a **carboxyl** group, and an **amino** group attached. The fourth group attached to the carbon atom is called the **R group.** Each of the 20 amino acids differs in its R group. R groups may be acidic, basic, polar, or nonpolar. The individual amino acids that make up a protein are linked together by **peptide bonds.** These bonds are formed by dehydration synthesis between the carboxyl group of one amino acid and the amino group of the next amino acid. Figure 9.5 illustrates the formation of a peptide bond between two amino acids by dehydration synthesis. A long chain of amino acids is called a **polypeptide chain.** The precise three-dimensional structure of a protein reflects both its amino acid sequence and the way in which the R groups of the individual amino acids interact with each other and with other molecules in their surrounding environment. Proteins play many important functions vital to an organism's survival. One of the most important groups of proteins is the enzymes. Enzymes are proteins that act as catalysts to speed up metabolic reactions within cells. They will be studied in more detail in a later lab.

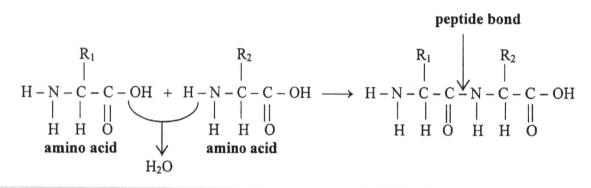

Figure 9.5: Formation of a Peptide Bond

Procedure: Chemical Tests to Identify Biological Molecules

In today's experiment, simple chemical tests will be performed to identify the presence of various types of biological molecules. These tests take advantage of the fact that different types of molecules have different functional groups and, therefore, different chemical properties. Thus, they will react to test **reagents** (also called **indicators**) differently. In each test, there will be one test tube that contains only water that will *not* react with the test reagent. This tube will serve as a **negative control** for the experiment by illustrating what a negative result for this test looks like. If this tube gives a positive result, something is wrong with the experiment. Each test will also include one test tube that is known to contain a particular substance that *will* react with the test reagent. This tube will serve as a **positive control** for the experiment by illustrating what a positive result looks like. If this tube gives a negative result, something is wrong with the experiment. Each test will also include a third tube that contains a solution of unknown composition. This tube represents the **experimental** solution. By comparing the appearance of the experimental tube with the positive and negative control tubes, it can be determined whether or not a particular substance is present in the unknown solution. The unknown may contain one or more different substances or none at all. The objective of the lab exercise is to determine the contents of the unknown solution.

When performing the chemical tests described below, the two control tubes will be done with a partner and an experimental tube will be tested by each individual student with his own unknown.

Step 1: EACH GROUP will obtain two unknown solutions from the instructor and **record the unknown number** in the space provided below the table on the data sheet.

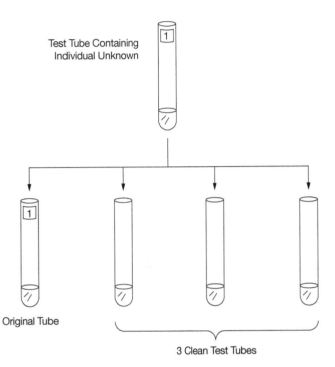

Test Tube Containing
Individual Unknown

Step 2: Immediately divide the unknown into four equal volumes into clean test tubes as shown at the right. One sample will be used for each of the four chemical tests performed.

Original Tube

3 Clean Test Tubes

Step 3: Place the four unknown tubes in the designated area of the test tube rack as shown below.

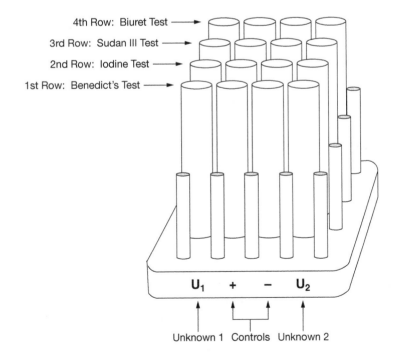

4th Row: Biuret Test
3rd Row: Sudan III Test
2nd Row: Iodine Test
1st Row: Benedict's Test

U₁   +   −   U₂

Unknown 1   Controls   Unknown 2

Test 1: The Benedict's Test for Reducing Sugars

All monosaccharides and some disaccharides have free aldehyde groups or ketone groups attached to their carbon skeleton. Sugars with this particular structure are called **reducing sugars** and can be detected by a simple chemical test known as the **Benedict's Test.** Not all sugars will give a positive result on this test. Only those that are reducing sugars, like glucose, will test positive. The Benedict's reagent is a bright blue color. When boiled, if a reducing sugar is present it changes progressively from blue to green, yellow, orange, and brick red, depending on the amount of reducing sugar present. Thus, Benedict's Test is not only a **qualitative** test, but also a **quantitative** test. It not only indicates *if* a reducing sugar is present, but also approximately *how much* is present.

Instructions (for each group):

One test tube represents the negative control, while another test tube represents the positive control.

1. Holding a ruler against the side of each tube, make marks at 1 cm and 2 cm from the bottom.

2. Fill the **negative control** tube to the 1-cm mark with **water.** Fill the **positive control** tube to the 1-cm mark with **glucose** solution.

3. To both tubes, add **Benedict's reagent** up to the 2-cm mark and mix thoroughly by swirling.

4. Now, each group will add an equal amount of **Benedict's reagent** to the **unknown** test tubes.

5. Use the test tube clamps provided to place the tubes in the boiling water bath. Leave the clamps on. Heat all tubes together for about 1 minute in the boiling water bath.

6. **CAREFULLY** remove the tubes using the test tube clamps and return to the test tube rack. Record the final colors of the solutions in the **data table.**

   Blue (no color change) = negative (-) for reducing sugar

   Green to yellow color = positive (+) for reducing sugar

   Yellow to orange color = positive (++) for reducing sugar

   Orange to red color = positive (+++) for reducing sugar

Questions:

1. Why is the first tube a negative control in this test?

2. Why is the second tube a positive control in this test?

3. Is reducing sugar present in the unknown? How was this determined?

Test 2: The Iodine Test for Starch

The **Iodine Test** can be used to test specifically for the presence of **starch.** Starch is a polymer of glucose units in which the chains are coiled up in a unique 3-dimensional shape. Dark golden-amber colored iodine interacts with this structure resulting in a distinctive blue-black color. Other carbohydrates, even those very similar to starch, lack the precise 3-dimensional shape to interact with the iodine molecules and thus do not give a positive result.

Instructions (for each group):

The first test tube represents the negative control, while the second test tube represents the positive control.

1. Holding a ruler against the side of each tube, make a mark at 1 cm from the bottom.

2. Fill the **negative control** tube to the 1-cm mark with **water.** Fill the **positive control** tube to the 1-cm mark with **starch** solution.

3. Now, add 5-10 drops of **iodine** solution to each of the control tubes AND to each **unknown** tube. Mix well by swirling.

4. Record the final colors of the solutions in the **data table.**

    Golden/amber color (no color change) = negative (-) for starch

    Blue/black color = positive (+) for starch

Questions:

1. Why is the first tube a negative control in this test?

2. Why is the second tube a positive control in this test?

3. Is starch present in the unknown? How was this determined?

Test 3A: The Sudan III Test for Lipids

The red dye, **Sudan III,** is a nonpolar liquid. Because of this, it will not mix readily with water, but instead, will separate from the water forming a red layer that floats on the surface. This is called a **hydrophobic interaction.** The Sudan III solution will mix readily with other nonpolar, hydrophobic molecules such as **lipids.** This difference in solubility provides the basis of the Sudan III Test for lipids.

Instructions (for each group):

The first test tube represents the negative control, while the second test tube represents the positive control.

1. Holding a ruler against the side of each tube, make a mark at 1 cm from the bottom.

2. Fill the **negative control** tube to the 1-cm mark with **water.** Fill the **positive control** tube to the 1-cm mark with **oil** solution.

3. Now, add 5-10 drops of **Sudan III dye** to each of the control tubes AND to each **unknown** tube. **MIX COMPLETELY by swirling.**

4. Note the appearance of the contents of each tube and record the results in the **data table.**

    Red layer floating on top of clear liquid = negative (-) for lipid

    Solid bright red color throughout = positive (+) for lipid

Questions:

1. Why is the first tube a negative control in this test?

2. Why is the second tube a positive control in this test?

3. Does the unknown contain lipid? How was this determined?

Test 3B: Filter Paper Test for Lipids

Add one drop each of distilled water on a filter paper in a petri dish; 1 drop of oil; and 1 drop of the unknowns separately. Label accordingly. Allow the drops to dry in air or by using a hair dryer.

Cover the paper with Sudan IV solution for three minutes. Rinse the paper with distilled water for one minute and observe the color of the circles.

A dark red spot indicates presence of lipids, and pale pink color on the rest of the paper indicates absence.

Test 4: The Biuret Test for Protein

The **Biuret** reagent is a pale blue color. However, if **protein** is present, the Biuret reagent will react specifically with the **peptide bonds** of the protein to produce a pale violet color. This change from blue to violet can best be observed if the tube is held in front of a white sheet of paper.

Instructions (for each group):

The white-coded test tube represents the negative control, while the green-coded test tube represents the positive control.

1. Holding a ruler against the side of each tube, make a mark at 1 cm from the bottom.
2. Fill the **negative control** tube to the 1-cm mark with **water.** Fill the **positive control** tube to the 1-cm mark with **protein** solution.
3. Now, add 10-15 drops of **Biuret reagent** to each of the control tubes AND to each **unknown** tube. Mix thoroughly by swirling.
4. Record the final color of each solution in the **data table.**

   Pale blue color = negative (-) for protein

   Pale violet color = positive (+) for protein

Questions:

1. Why is the first tube a negative control in this test?

2. Why is the second tube a positive control in this test?

3. Does the unknown contain protein? How was this determined?

Conclusions:

At this point, examine the results of the four tests that have been performed on the unknown. Based on these test results, state the composition of the unknown solution in the space provided on the data sheet.

Clean-Up:

1. Empty the contents of the tubes as instructed in the waste bottles and place all of the used tubes in the "Used Test Tubes" bin provided.

Summary of Procedure

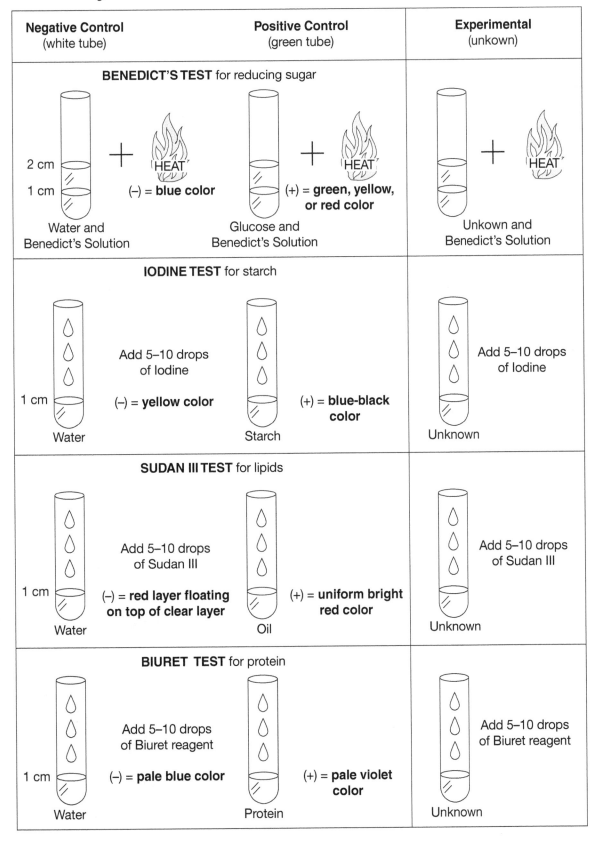

Negative Control (white tube)	Positive Control (green tube)	Experimental (unkown)
BENEDICT'S TEST for reducing sugar		
2 cm / 1 cm + HEAT (–) = **blue color** Water and Benedict's Solution	+ HEAT (+) = **green, yellow, or red color** Glucose and Benedict's Solution	+ HEAT Unkown and Benedict's Solution
IODINE TEST for starch		
Add 5–10 drops of Iodine (–) = **yellow color** 1 cm Water	Add 5–10 drops (+) = **blue-black color** Starch	Add 5–10 drops of Iodine Unknown
SUDAN III TEST for lipids		
Add 5–10 drops of Sudan III (–) = **red layer floating on top of clear layer** 1 cm Water	(+) = **uniform bright red color** Oil	Add 5–10 drops of Sudan III Unknown
BIURET TEST for protein		
Add 5–10 drops of Biuret reagent (–) = **pale blue color** 1 cm Water	(+) = **pale violet color** Protein	Add 5–10 drops of Biuret reagent Unknown

Questions for Study and Review

1. Answer the following questions about the Benedict's Test.

    a. What color is the Benedict's reagent?

    b. What type of organic molecules does it test for?

    c. What color(s) indicate a positive test?

    d. Why is the Benedict's test both a qualitative and a quantitative test?

2. Answer the following questions about the Iodine Test.

    a. What color is iodine solution?

    b. What type of organic molecules does it test for?

    c. What color indicates a positive test?

3. Answer the following questions about the Sudan III Test.

    a. What color is the Sudan III solution?

    b. Is the Sudan III polar (hydrophilic) or nonpolar (hydrophobic)?

    c. What type of organic molecules does it test for?

    d. What does a positive Sudan III test look like?

    e. What does a negative Sudan III test look like and why does it look this way?

4. Answer the following questions about the Biuret Test.

   a. What color is the Biuret solution?

   b. What type of organic molecules does it test for?

   c. What does a positive Biuret test look like?

5. What are the 4 main elements found in living things?

6. What is the difference between an organic and an inorganic molecule?

7. Name the four major classes of organic macromolecules in living things.

8. What are functional groups and why are they important? Name 6 examples and give their formulas.

9. Name the three major types of carbohydrates. Give an example of each type.

10. What one property do all lipids share in common?

11. How many different amino acids are found in the proteins of living things?

12. What is a peptide bond?

13. Fill in the following chart:

Test	Positive Control	Positive Result	Negative Control	Negative Result
	starch solution			
		pale violet		
Benedict's			water	
				red layer on top

Name: _____   Date: _____

Biological Molecules Data Sheet

Data Table: Results from Chemical Tests

Test 1: *Benedict's Test* for Reducing Sugar

Substance Tested	Final Color	Results (+ or –)	Reducing Sugar Present?
Water			
Glucose Solution			
Unknown Solution 1			
Unknown Solution 2			

Test 2: *Iodine Test* for Starch

Substance Tested	Final Color	Results (+ or –)	Starch Present?
Water			
Starch Solution			
Unknown Solution 1			
Unknown Solution 2			

Test 3: *Sudan III Test* for Lipid

Substance Tested	Final Appearance	Results (+ or –)	Lipid Present?
Water			
Oil Solution			
Unknown Solution 1			
Unknown Solution 2			

Test 4: *Biuret Test* for Protein

Substance tested	Final Color	Results (+ or –)	Protein Present?
Water			
Protein Solution			
Unknown Solution 1			
Unknown Solution 2			

Conclusions About the Unknown:

1. My **unknown numbers** were _____ and _____.

2. Based on the results of testing recorded above, my unknown contains the following substance(s):

   Unknown _____ Unknown _____

10

Enzymes:
Biological Catalysts

Introduction

A **catalyst** is any substance that speeds up the rate of a chemical reaction without being permanently changed or consumed in the reaction. Catalysts lower the amount of activation energy needed to get a chemical reaction started. Because less activation energy is required when a catalyst is present, more molecules possess the energy necessary to overcome this "energy barrier," and the reaction occurs more rapidly.

A catalyst may be inorganic or organic. In cells, metabolism is controlled by organic catalysts called **enzymes.** Most enzymes are protein molecules generally composed of several hundred or more amino acids joined in a specific sequence. In order to become activated, some enzymes require a non-protein component called a **cofactor.** Cofactors may be inorganic metal **ions** (such as Ca^{+2}) from *minerals* or they may be small organic molecules called **coenzymes** from *vitamins.*

In an enzyme-catalyzed reaction, the molecule that is acted upon by the enzyme called its **substrate.** In order for the enzyme to affect the substrate, it must form a physical attachment to the substrate. The specific location on the enzyme where the substrate binds is called the **active site.** When the enzyme attaches to the substrate, it forms the **enzyme-substrate complex.** Once this complex forms, the enzyme catalyzes the reaction, and the **products** of the reaction are released. (See Figure 10.1 on next page.) As is true for all catalysts, the enzyme is unchanged in this process and can immediately bind to another substrate molecule and catalyze the same reaction over and over again. A single enzyme molecule can carry out thousands of reaction cycles every second!

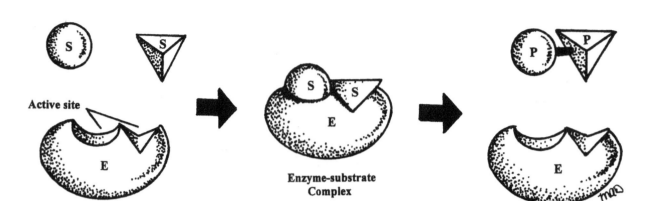

Figure 10.1:   Enzyme Action

Enzymes are extremely specific for the type of reaction they will catalyze. Each particular enzyme will only catalyze a specific type of reaction or a group of closely related reactions. The basis for enzyme specificity resides in its **three-dimensional shape.** Enzymes are globular proteins with complex shapes determined by the specific type, number, and sequence of amino acids that compose the enzyme. The **active site** is usually a groove or cleft in the enzyme molecule where the substrate can "dock" into place. There must be an approximate match between the shape of the substrate and that of the active site to form an enzyme-substrate complex. The binding of the substrate to the active site induces a more precise fit. Molecules that are not the correct shape cannot bind to the enzyme and cannot serve as substrates.

If a molecule other than the substrate binds to the active site of an enzyme, it prevents the substrate from attaching, thereby inactivating the enzyme. Such a molecule is called an **enzyme inhibitor.** Some inhibitors have reversible effects, but some are irreversible. The deadly poisons cyanide and nerve gas are irreversible inhibitors of enzymes that are critical to the cell's metabolism.

Under certain circumstances, the shape of an enzyme's active site may be altered. This can have a negative effect on the enzyme's ability to bind its substrate and catalyze the reaction. Alteration of the three dimensional shape of a protein is known as **denaturation.** Conditions that may cause denaturation include exposure to extremes of temperature and pH, electricity, radiation, and various chemicals. When exposed to these types of conditions, enzymes quickly denature and the shapes of their active sites are altered, resulting in their inactivation. If the conditions are not too extreme and return to normal quickly, denaturation may be reversible. Most of the time, however, denaturation results in a permanent loss of enzyme activity. These conditions must be carefully controlled in living organisms (and in laboratory experiments) so that enzymes work at peak efficiency.

Many different factors affect enzyme activity and the rate of an enzyme-catalyzed reaction. These include substrate concentration, enzyme concentration, temperature, pH, and presence of chemical regulators. The effect of each of these factors is discussed below.

1. **Substrate concentration:** As substrate concentration increases, the rate of the reaction increases until the enzyme becomes saturated with substrate. At this point, the rate levels off and the reaction is occurring at its maximum velocity (V_{max}).

2. **Enzyme concentration:** As the concentration of enzyme is increased, the rate of the reaction increases. As long as there is an unlimited amount of substrate present, the rate will continue to increase as more enzyme is added.

3. **Temperature:** All chemical reactions speed up as temperature is increased. As temperature increases, reacting molecules collide more violently and more of them have sufficient kinetic energy to overcome the energy barrier and react. Reactions catalyzed by enzymes also speed up as temperature increases. However, most enzymes have an optimum temperature at which they reach their maximum rate of activity. As temperature is increased above optimum, the structure of the enzyme starts to be disrupted. As the enzyme denatures, the rate of activity drops rapidly. The positive effect of speeding up the reaction is negated by the negative effect of denaturing the enzymes. Many proteins are denatured by temperatures around 40–50°C, but some are still active at 70–80°C, and a few can even withstand boiling.

4. **pH:** Each enzyme has an optimum pH at which it works best. As the pH is increased or decreased beyond the optimum range, the enzyme's shape is disrupted and activity decreases. Most enzymes have an optimum pH near neutral and become denatured in extremely acidic or alkaline environments. Some enzymes, such as those that act in the human stomach where the pH is very low, have an appropriately low optimum pH.

5. **Allosteric Regulators:** Allosteric regulators are small molecules, other than the substrate, that may temporarily bind to an enzyme, turning it "on" or "off' by changing the shape of its active site. Regulators that cause the enzyme to become active are called **activators,** while those that cause the enzyme to become inactive are called **inhibitors.** Because regulators can activate or inactivate enzymes, cells can use them to regulate metabolism.

The Enzymatic Activity of Salivary Amylase

Background Information

The saliva of most people contains a digestive enzyme known as **salivary amylase.** This enzyme acts upon the substrate starch and **hydrolyzes** it (breaks it down) into simpler disaccharides and monosaccharides, including reducing sugars such as maltose and glucose. This reaction is given below:

$$\text{starch} \xrightarrow{\text{salivary amylase}} \text{maltose and glucose (reducing sugars)}$$

Like most enzymes found in humans, this enzyme has maximum activity at the optimum temperature of 37°C (body temperature) and a pH of 10.4. Students will collect and use saliva as a source of the enzyme amylase. (Note: There are a few individuals who actually lack this enzyme in their saliva due to a genetic mutation!)

Purpose

In this experiment, the catalytic activity of amylase will be investigated at its optimum temperature and pH. The effects of increased temperature and decreased pH on the enzyme's activity will also be determined.

Procedure

In this experiment, the hydrolysis (breakdown) of starch will be assayed by performing the Benedict's Test to detect the presence of reducing sugars. If starch *is* hydrolyzed, the reducing sugars present will cause a *positive* Benedict's Test. If starch is *not* hydrolyzed, the absence of reducing sugars will result in a *negative* Benedict's Test. Based upon these results, one can determine if the amylase enzyme was active.

To avoid introducing a random variable, *one* student in each lab group should contribute *all* of the saliva. Since any residue in the mouth from food or drink could interfere with the results of the experiment by creating false positive tests for reducing sugar, the mouth must be **thoroughly** rinsed with water before collecting the saliva.

1. Obtain 5 clean test tubes. Holding a ruler against the clear glass of each tube, make marks at 1 cm and 2 cm from the bottom.

2. Place the following in the test tubes as indicated below:

   Tube 1: saliva to the 1-cm mark + water to the 2-cm mark
   Tube 2: starch to the 1-cm mark + water to the 2-cm mark
   Tube 3: saliva to the 1-cm mark + starch to the 2-cm mark
   Tube 4: saliva to the 1-cm mark (boiled **first** for 10 minutes in the hot water bath and cooled) + starch to the 2-cm mark
   Boil the saliva *before* adding the starch solution!
   Tube 5: saliva to the 1-cm mark + 15 drops HCl (mixed thoroughly **first**) + starch to the 2-cm mark
   Mix the HCl with the saliva *before* adding the starch solution!

3. Now place all 5 tubes in a 37°C warm water bath and let them incubate for 5 minutes.

4. Remove the tubes from the warm water bath and return them to the test tube rack.

   To each of the 5 tubes add a volume of Benedict's solution equal to that of the liquid in each of the 5 tubes. Immerse all the tubes in the boiling water bath and watch for a color change. (Recall that Benedict's solution is an indicator for reducing sugars, like glucose and maltose. It first turns green, then yellow, then orange, and finally brick red in the presence of increasing quantities of reducing sugar.)

5. Record the test results.

6. Place all used test tubes in the "dirty test tube" bin.

Summary of Procedure

Individual Set-ups

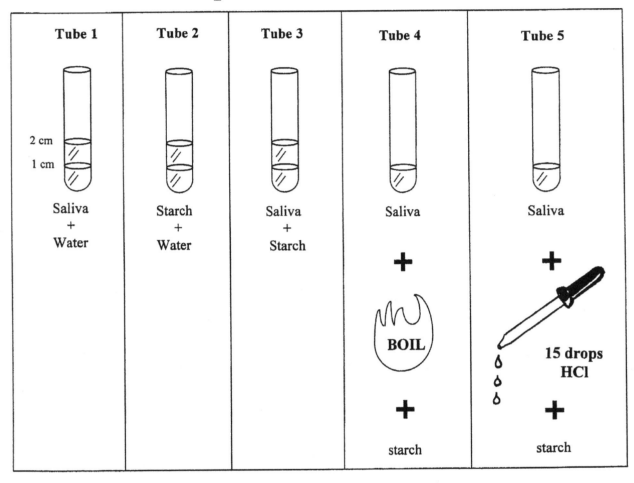

Tube 1	Tube 2	Tube 3	Tube 4	Tube 5
2 cm 1 cm Saliva + Water	Starch + Water	Saliva + Starch	Saliva + BOIL + starch	Saliva + 15 drops HCl + starch

Procedure for All 5 Tubes

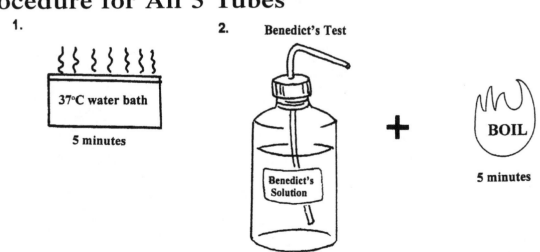

1.
37°C water bath
5 minutes

2. Benedict's Test
Benedict's Solution
+
BOIL
5 minutes
Add to each tube.

EXERCISE 10

Questions for Study and Review

1. Define catalyst.

2. Are all enzymes catalysts? Are all catalysts enzymes? Explain.

3. The graph below illustrates the effect of temperature on the activity of both an inorganic and an organic catalyst.

   a. Which line represents the inorganic catalyst and which line represents the organic catalyst?

   b. Why is there a sharp drop at the point indicated by the arrow?

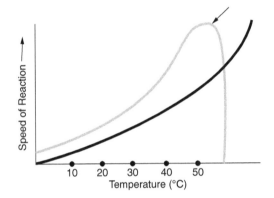

4. What are coenzymes and how are they obtained?

5. What is denaturation and how does it affect enzyme activity?

6. Define the terms: substrate, active site, enzyme-substrate complex.

7. What is the basis of enzyme specificity for substrate?

8. How does substrate concentration affect enzyme activity?

9. How does enzyme concentration affect enzyme activity?

10. How does temperature affect enzyme activity? What would be the optimum temperature for most human enzymes?

11. How does pH affect enzyme activity? What would be the optimum pH for most human enzymes?

12. What are allosteric regulators? Name two types.

Name: _____   Date: _____

EXERCISE 10

The Enzymatic Activity
of Salivary Amylase

1. Complete the chart below:

Tube	Contents	Resulting Color	Reducing Sugar (Present or Absent)	Starch Hydrolysis (Yes or No)
Tube 1	Saliva only			
Tube 2	Starch only			
Tube 3	Saliva + Starch			
Tube 4	Boiled Saliva + Starch			
Tube 5	HCl + Saliva + Starch			

*Using the color key on page 68, plot a suitable graph of your results on a graph paper and explain your findings.

2. What was the **name** of the enzyme studied in this lab? _____

   What was its **substrate?** _____

3. Why was the temperature of the warm water bath set at 37° C? (HINT: Consider the *source* of the enzyme.) _____

   _____

4. Write the word equation for starch hydrolysis. Write the name of the enzyme for this reaction above the arrow.

   What does Benedict's solution test for that would show whether or not this reaction occurred?

   _____

5. Considering the content(s) of each tube, **explain *why* starch hydrolysis did or did not occur** in that particular tube.

   a. Tube 1: _____

   b. Tube 2: _____

85

c. Tube 3: _____

d. Tube 4: _____

e. Tube 5: _____

6. Which two tubes acted as negative controls? _____

7. Which tube acted as a positive control?_____

8. Which two tubes were experimental tubes **and** what variable was each testing?

_____

_____

9. Offer a **possible explanation** for each of the following situations:

a. A false positive result in tube 1: _____

b. A negative result in tube 1 but a false positive result in tube 4: _____

_____

c. A negative result in tube 1 but a false positive result in tube 5: _____

_____

d. A negative result in tube 3: _____

_____

11

Cells

Introduction

All organisms are composed of the basic unit of life called the cell. Cells vary in their shape and size, yet share certain basic features in their design and in the way they function (Figure 11.1). Each cell is a functional unit capable of carrying on all of the processes associated with life. As you compare one cell type to another in today's lab, keep in mind that you will be observing cells at different levels of organization and complexity. You will be asked to compare the features of cells, as they relate to the life processes of organisms. (See Figure 11.1).

After completing this exercise, you should be able to:

1. List the differences between prokaryotic and eukaryotic cells.

2. Use the compound microscope to view cells and their structures.

3. Describe and define the structures that you expect to see in animal and plant cells.

Prokaryotic Cells

Bacteria and the **cyanobacteria** (blue-green algae) are said to be **prokaryotes** because they do not contain a membrane-bound nucleus or any other membrane-bound organelles. Also, the DNA found in these cells is not organized into chromosomes as in the cells of more complex organisms. These organisms also possess a cell wall, sometimes covered by a gelatinous capsule.

Bacteria

Bacteria are examples of prokaryotic cells. They vary in size from 1–10 μm and have various shapes and arrangements (Figure 11.2). You will be observing living bacteria that are used in a commercial process that converts milk to yogurt. The species is called *Lactobacillus acidophilus* and is present in yogurt that contains **active cultures** of the organisms. Also, you will be comparing the sizes and shapes of other species of bacteria using prepared slides.

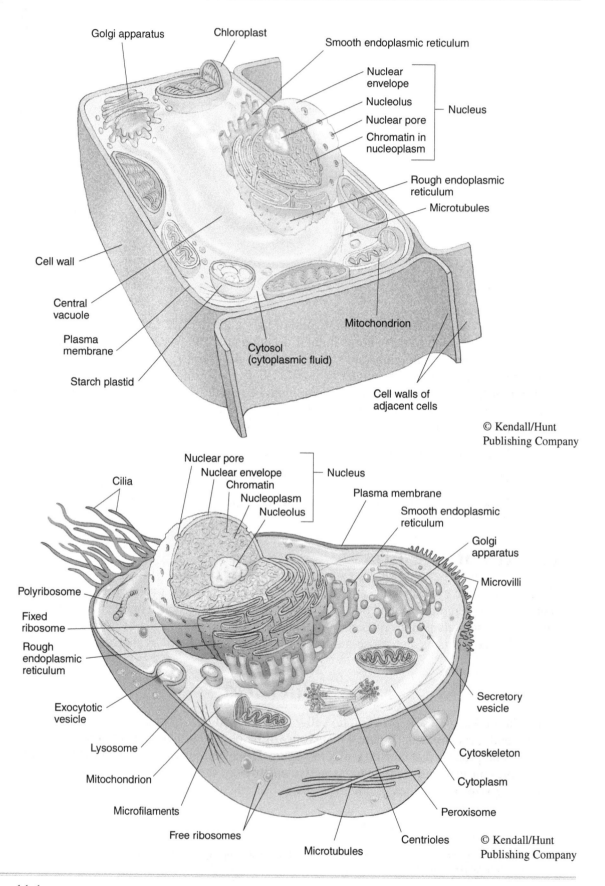

Golgi apparatus   Chloroplast   Smooth endoplasmic reticulum

Nuclear
envelope
Nucleolus   — Nucleus
Nuclear pore
Chromatin in
nucleoplasm

Rough endoplasmic
reticulum

Microtubules

Cell wall

Central
vacuole

Plasma
membrane

Starch plastid

Cytosol
(cytoplasmic fluid)

Mitochondrion

Cell walls of
adjacent cells

© Kendall/Hunt
Publishing Company

Cilia

Nuclear pore
Nuclear envelope   — Nucleus
Chromatin
Nucleoplasm
Nucleolus

Plasma membrane

Smooth endoplasmic
reticulum

Golgi
apparatus

Microvilli

Polyribosome

Fixed
ribosome

Rough
endoplasmic
reticulum

Exocytotic
vesicle

Lysosome

Mitochondrion

Microfilaments

Free ribosomes

Microtubules

Centrioles

Secretory
vesicle

Cytoskeleton

Cytoplasm

Peroxisome

© Kendall/Hunt
Publishing Company

Figure 11.1

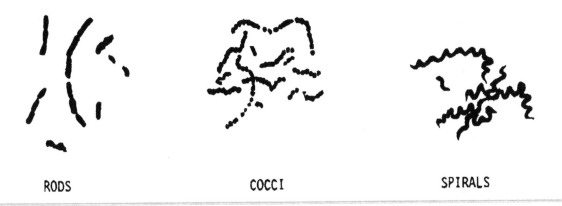

RODS   COCCI   SPIRALS

Figure 11.2

From *Laboratory Guide to Human Biology, Online Version* by Robert Amitrano and Martha Lowe. Kendall/Hunt Publishing Company.

Procedure

1. Obtain a clean microscope slide and a coverslip.

2. Using a toothpick, place a tiny dab of yogurt on the slide, and mix the yogurt with a drop of water.

3. Add a coverslip and examine the yogurt with a compound microscope.

4. Some secrets to success:
   a. Reduce the light, using the iris diaphragm, or drop the condenser to a lower spot.
   b. Focus with the low-power objective (10×), then rotate the high-power objective into place, and refocus with the fine adjustment knob.

5. The bacteria appear as tiny rod-shaped structures, and are found between the "pieces" of yogurt.

6. Make a sketch of the cells below or from micrographs provided: Magnification—400×

7. Now, obtain a prepared slide of different types of bacterial cells. Observe these using the microscope and draw them in the space provided below:

   Magnification—400×

From *Biological Investigations*, Revised Printing by Gayne Bablanian. Copyright © 2002 by Gayne Bablanian. Used with permission of Kendall/Hunt Publishing Company.

8. Using Figure 11.2 as a guide, list the different shapes of bacteria that you observed.

Eukaryotic Cells

Eukaryotic cells are much larger than prokaryotic cells, and have a membrane-bound **nucleus** that contains chromosomes made of DNA and protein. Also, the cytoplasm contains membrane-bound organelles that are specialized structures each with a specialized function. Eukaryotic cells are found in the kingdoms **Protista, Fungi, Plantae** and **Animalia.**

Plants

Plant cells are more complex and possess several additional cytoplasmic organelles. The outer covering of the cell is a rigid **cell wall** composed of **cellulose.** The cell wall overlies the plasma membrane. It protects and supports the cell while not interfering with the movement of molecules across the cell membrane.

In the cells of the pond plant, *Elodea,* you will observe colored organelles (usually green) called **chloroplasts** (Figure 11.3). Chloroplasts are the site for the process of photosynthesis. The nucleus in these cells will not be obvious because it is pushed to one side by the large, fluid-filled, **vacuole.** Be sure to look for **cyclosis** or cytoplasmic streaming. The chloroplasts may be moving around the inside of the cell membrane.

Also, you will be observing plant cell structure using the red onion. The onion is the root part of the plant. Therefore, photosynthesis does not occur in this portion of the plant. You will not find any chloroplasts in these cells.

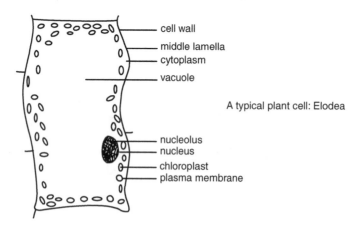

cell wall
middle lamella
cytoplasm
vacuole

A typical plant cell: Elodea

nucleolus
nucleus
chloroplast
plasma membrane

Figure 11.3

Procedure

1. Obtain a sprig of *Elodea* and tear off a leaf from the top part of the sprig.

2. Make a wet mount of the leaf. If you place the top surface facing up, it will be easier to see the larger cells that are located on the top surface of the leaf.

3. Examine the leaf with your microscope. How many layers of the cells are present?

   _____

4. Under low or high power, focus on one cell. Each of the small, rectangular "boxes" represents a cell surrounded by the cell wall.

5. Chloroplasts appear as small green bodies within the cell. Observe the cell for at least a minute to see the phenomenon called cyclosis.

6. Draw the *Elodea* cell below:

   Low Power (100×)   High Power (400×)

7. List the organelles and structures that are visible to you. (Hint: There should be at least four items in your list.)

   _____   _____

   _____   _____

8. Describe the process of cyclosis in your own words.

   _____

   _____

9. Obtain a piece of red onion. Snap the leaf backwards and remove the thin layer of epidermis formed at the break point (see Figure 11.4).

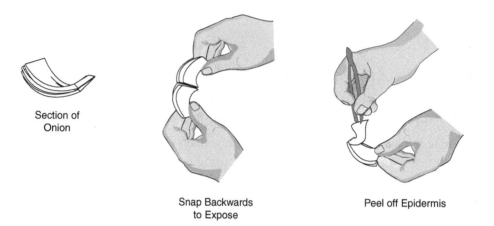

Section of
Onion

Snap Backwards
to Expose

Peel off Epidermis

Figure 11.4

10. Gently spread this thin tissue in a drop of water on a microscope slide. Add a coverslip and observe the tissue.

11. Put a drop of iodine at one edge of the coverslip. Touch the fluid at the opposite edge of the coverslip with a small piece of paper towel. This will draw the iodine under the coverslip.

12. Examine the onion cells under low and high power, and draw below.

Low Power (100×)   High Power (400×)

13. List the organelles and structures that are visible to you. (Note: You should be able to see a round shaped body inside the nucleus. This is called the **nucleolus.**)

_____

_____

14. How do the two plant cells that you have examined differ in structure?

_____

_____

Animals

Animal cells vary greatly in size, shape and function. Unlike plant cells, animal cells **do not** have a cell wall, chloroplasts, or vacuoles. The plasma membrane protects the cell, and allows for more flexibility in the overall structure of the animal body. A readily available type of animal cell is the human epithelial cell found inside the mouth, on the inner surface of the cheek (Figure 11.5).

Procedure

1. Place a small drop of water on a microscope slide.
2. Using the flat edge of a clean toothpick, gently scrape the inner surface of your cheek.
3. Mix the "scraping" into the water. Discard the toothpick into the RED BIOHAZARD CONTAINER.
4. Using a toothpick add a small drop of methylene blue stain and add a coverslip.

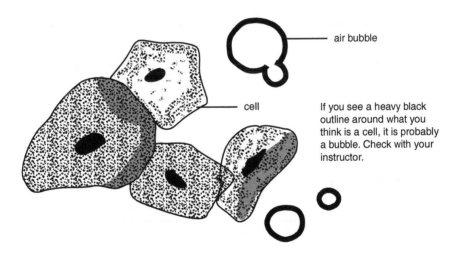

air bubble

cell

If you see a heavy black outline around what you think is a cell, it is probably a bubble. Check with your instructor.

Figure 11.5

5. The cheek cells are flat and sometimes are seen in clusters. Under low power, search for bluish cells with small darker blue areas in the center.

6. Switch to high power and draw several cells below: (400×). Label the nucleus, plasma membrane, and the cytoplasm.

Comparison of Bacterial, Animal and Plant Cells

In the table below, indicate which structures and organelles are PRESENT or ABSENT for each category of organism. Use your textbook or any other information to complete this exercise.

	Bacterium	*Plant*	*Animal*
Cell Wall			
Plasma Membrane			
Flagella			
ER			
Ribosomes			
Golgi Apparatus			
Nucleus			
Chloroplasts			
Mitochondria			
Chromosomes			
Vacuoles			

Cell Structures

In the table that follows, give a description of each eukaryotic cell structure and its function in the cell.

Structure	Description	Function
Cell Wall		
Plasma Membrane		
Flagella		
Cytoskeleton		
ER		
Ribosomes		
Golgi Apparatus		
Nucleus		
Nucleolus		
Chromosomes		
Mitochondria		
Chloroplasts		
Lysosomes		

Diffusion and Osmosis

12

A *solution* is a homogeneous molecular dispersion of two or more substances. Typically, a solution is formed by dissolving a solid (the *solute*) in a liquid (the *solvent*). For example, in a sugar solution, sugar is the solute and water is the solvent. The movement of water and dissolved substances involves various physical processes. The energy necessary for this movement is derived from molecular motion. Molecular motion within and between cells can be explained by the laws of physics and chemistry. One of these laws, the law of *diffusion,* is of particular importance to our understanding of the movement of molecules into and out of cells. Diffusion occurs without requiring the expenditure of cellular energy. In a given quantity of matter, the amount of free energy available for work (the movement of matter) is defined as free energy. The energy available to move matter is defined as the chemical energy of the matter. Chemical energy is proportional to concentration—a relationship that allows us to more accurately state the law of diffusion. The law of diffusion states that molecules tend to move from areas of higher chemical potential (and higher concentration) to areas of lower chemical potential (and lower concentration).

Osmosis, a special type of diffusion, is the movement of **WATER** molecules from areas of higher water potential (lower solute concentration to regions of lower water potential (higher solute concentration) across a **SELECTIVELY PERMEABLE MEMBRANE.** A selectively permeable membrane is a membrane permeable to water and small solute molecules but impermeable to large solute molecules. In living cells, the cell membrane is selectively permeable, but for this experiment, we will use dialysis tubing to construct our "cells". Like cell membranes, dialysis tubing has pores which select for molecular size. Small molecules such as water can freely pass through the membrane while larger molecules such as starch are unable to pass through the membrane.

Solvent (water) moving through selectively permeable membrane always moves from the solution that contains a lower solute concentration to the solution containing a higher solute concentration. The solution containing the lower solute concentration is said to be *hypotonic* (hypo = less than) and the solution with the higher solute concentration is *hypertonic* (hyper = more than) to the solution with the lower solute concentration. If the two solutions across a selectively permeable membrane contain equal solute concentrations, they are said to be *isotonic* and there will be no net movement of water across the membrane.

The difference in concentration of like molecules in two regions is called a *concentration gradient.* Diffusion and osmosis take place **DOWN** concentration gradients. With time, the concentration of solvent and solute molecules becomes equally distributed, the gradient ceases to exist. If the gradient ceases to exist, the system is said to be at *equilibrium.* However, you should keep in mind, even at equilibrium, molecules are always in motion. Thus solvent and solute molecules continue to move because of randomly colliding molecules. At equilibrium there is no *net change* in the concentration.

Exercise I. Brownian Movement

To understand the movement of substances through membranes, it is important to realize that molecules are usually in constant motion. Molecular motion is a form of energy—the kinetic energy of molecules. Although it is impossible to see individual molecules, their movement is manifested by the phenomenon known as *Brownian movement.* This is the movement of small particles caused by the bombardment of the particles by millions of water molecules. This movement continues indefinitely as long as the water does not evaporate. Because the movement of water molecules is random, this bombardment is not equal on all sides and the particles appear to jiggle in an irregular fashion.

Objective:

Observe the phenomenon of Brownian movement and understand its role in the movement of molecules.

Materials:

1. Bon Ami scouring powder, carmine red and Indian ink.
2. Microscope slides and coverslips.
3. Dropper bottles with distilled water.
4. Spatulas.
5. Microscopes.

Procedure:

1. Place a small portion of Carmine red/Indian ink or on a microscope slide, add several drops of distilled water, and add a coverslip.

2. Place the slide on the microscope stage and once the water currents have ceased observe the movement of the tiny particles under high power.

3. Follow a single particle. How would you describe its motion? Is its movement continuous or does it speed up and slow down? _____

   _____

   _____

4. Place your light source close to your microscope for several minutes. This will warm the solution slightly. Does the heat affect the movement? _____

   _____

   _____

   How? _____

   _____

5. Can you see water molecules? _____ Are the water molecules moving? _____ What is the evidence that supports your answer?

_____

_____

6. How is this movement related to the following diffusion and osmosis experiments?

_____

_____

_____

Exercise II. Diffusion: Liquids Through a Colloid (demonstration)

Surely you have smelled perfume or after-shave (or worse smells) wafting from someone on the other side of a room. This is an example of diffusion at work; the movement of molecules from areas of higher chemical potential (and higher concentration) to areas of lower chemical potential (and lower concentration). During this process, molecules of one substance pass through and freely intermingle with those of another. Remember, diffusion is based upon the fundamental principle that all molecules—solids, liquids, and gases—are in constant motion.

In this demonstration you will observe the diffusion of two liquids through a colloid. A colloid system occurs when a finely divided solid is suspended in a liquid. In this instance, the colloid system we will use consists of a 2% agar gel.

Objectives:

1. Understand the process of diffusion.

2. Observe this example of diffusion; the movement of two liquids through a colloid system.

Materials:

1. Petri dishes containing 2% agar (5 mm deep).

2. Cork borer.

3. Dropper bottle with a dilute aqueous solution of potassium ferricyanide.

4. Dropper bottle with a dilute aqueous solution of ferrous sulfate.

Procedure:

1. The laboratory instructor will prepare this demonstration prior to the start of lab.

2. Obtain Petri dish containing 2% agar 5 mm deep. Using a cork borer, cut two holes into the agar approximately 1 cm apart.

3. Place the Petri dish on a level surface and fill one well with a dilute aqueous solution of potassium ferricyanide taking care not to overfill.

4. Fill the other well with a dilute aqueous solution of ferrous sulfate. Again, take care to not overfill the well.

5. When solutions of potassium ferricyanide and ferrous sulfate meet, an intense blue precipitate *(Prussian blue)* forms. Observe the plate at regular intervals for the formation of a thin blue line midway between the wells. Prussian blue will form as the solutions diffuse to meet each other.

6. Observe at 15 minute intervals until the end of lab. Allow the petri dish to remain overnight for the next laboratory period. Compare your results with earlier labs.

7. Draw and discuss the results of this demonstration.

Exercise III. Osmosis in a Nonliving System

Osmosis is the movement (diffusion) of water molecules from areas of higher water potential to regions of lower water potential across a selectively permeable membrane. In this exercise dialysis tubing will serve as the selectively permeable membrane. In effect, the dialysis tubing will simulate the plasma membrane of a living cell, the contents represent the cell contents, and the surrounding solution represents the prevailing environmental conditions.

Objectives:

1. Understand the process of osmosis and distinguish between osmosis and diffusion.

2. Determine the direction of the net movement of water for the experimental systems in this exercise.

3. Understand the use of the words hypertonic, hypotonic, and isotonic for the experimental systems in this exercise.

Materials:

1. Approximately 30 presoaked 15-cm lengths of 44-mm flat diameter dialysis tubing per lab.

2. Twine.

3. 1% starch suspension.

4. Iodine (I_2KI).

5. 25% sucrose (colored red).

6. Pure water.

7. Pure water (colored red).

8. Pipettes.

9. Balances.

10. Wax pencils.

Procedure:

1. Work in groups of four.

2. Each group should obtain five 15-cm lengths of 44-cm flat diameter dialysis tubing.

3. Roll the tubing briskly between your fingers and thumb to separate the two layers. Run water through to open the tube full length.

4. Seal off one end of the dialysis tubing by looping it over itself to tie a knot in the tube, making a bag with one open end.

5. Fill the dialysis tubing about two-thirds full with the appropriate experimental solution. Close the free end of the tube as directed. Remove as many air bubbles as possible before the final tying.

6. Work quickly, so that the dialysis tubing does not dry out. After the dialysis bag has been prepared, rinse the outside of the bag with distilled water to wash away any spilled solution.

7. Repeat this procedure until you have five dialysis bags that contain the solutions indicated in the table.

8. Weigh each bag by placing your filled dialysis bag in the weighing boat and record the weight in the table provided.

9. Place each filled dialysis bag in a small empty beaker and pour just enough of the solution indicated by the chart into the beaker to cover the bag. Observe the solution or the contents of the bag for any color change.

10. After approximately one hour examine and weigh all bags, record the weights, and supply any information called for in the table on the next page.

11. Did the bags gain weight, lose weight, or stay the same weight? Why? For each of the dialysis bags, in which direction did water move? Why? Indicate whether the solution surrounding each dialysis bag was hypertonic, hypotonic, or isotonic in comparison to the bag's contents.

Osmotic Behavior

Bag Number	Contents of Sausage (bag)	Original Weight at Start (in grams)	To Be Immersed In	Weight after One Hour	Appearance after One Hour (add sketch)
1.	1% starch suspension		iodine (I_2KI)	1.	
2.	Iodine (I_2KI)		1% starch suspension	2.	
3.	25% sucrose (colored red)		Pure water	3.	
4.	Pure water		25% sucrose	4.	
5.	Pure water (colored red)		Pure water	5.	

Draw a suitable graph on a graph paper and explain your results.

Exercise IV. Osmosis in Living Systems

When osmosis occurs within living systems, cells can either gain or lose water. This gain or loss can have serious consequences for the survival of an organism. Just as we have already seen, the direction of net water movement is determined by the solute concentration inside and out of a cell. *Plasmolysis* is defined as the separation of plant protoplasts from cell walls because of the loss of water from protoplasts by osmosis. This loss occurs because cells are surrounded by a solution of higher solute concentration (a hypertonic solution). When plant cells are firm due to water uptake, the tissue is referred to as *turgid*. We will examine both of these phenomena.

Objectives:

1. Observe plasmolysis and turgidity of plant cells and tissues.

2. Understand the underlying mechanisms that control the movement of water into and out of plant cells.

Materials:

1. *Elodea* plants and Potato tubers.

2. Microscope slides and coverslips.

3. Dropper bottles of distilled water.

4. Dropper bottles of a 10% NaCl solution.

5. Microscopes.

6. Cork borers.

7. Petri dishes.

Procedure:

1. **Plasmolysis:** Work in pairs. Prepare a wet mount of an *Elodea* leaf using distilled water.

2. Examine the leaf using your microscope and draw this leaf in the space provided.

3. Place a drop of 10% NaCl at the edge of the coverslip and draw it under by absorbing water on the opposite side using a piece of filter paper.

4. Quickly examine the cells under the microscope and note any change in the cells contents. Cells near the margin of the leaf should react first.

5. Draw a cell, in the space provided, that has undergone plasmolysis. What does this cell look like? What has taken place in the cell to account for this change?

   _____

   _____

6. Draw distilled water under the coverslip and observe the cells. What has taken place in the cell to account for this change?

   _____

Normal *Elodea* Leaf   Plasmolyzed *Elodea* Leaf

7. Work in pairs. Using a cork borer cut two pieces of potato tuber exactly the same size. They should be about 15–20 mm long.

8. Place one piece in a Petri dish containing 10% NaCl, and the other in a Petri dish with distilled water.

9. After three minutes, and every few minutes thereafter, compare the flexibility of the two pieces. Be sure not to cross-contaminate the two solutions. _____

   _____

   _____

   _____

   _____

10. What is happening in the cells of the potato that may account for the change in flexibility?

   _____

11. After you note a marked difference in the flexibility of the two pieces, wash the piece from the 10% NaCl solution with distilled water, and place it in the Petri dish containing distilled water.

12. Continue to compare the two potato pieces at regular intervals. What are your observations? How do you account for any changes?

   _____

   _____

   _____

Photosynthesis and the Leaf

The basic source of energy for all life on the Earth is the light of the Sun. Light energy is used by green plants to convert inorganic compounds to complex organic molecules by the process of *photosynthesis*. Photosynthesis may be the most important biological process sustaining life. By trapping the Sun's energy in glucose and eventually starch, photosynthesis provides virtually all of the energy used by organisms. This is the only process by which chemical energy is added to the global ecosystem. Additionally, all of the oxygen that we breathe is released as a byproduct of the photosynthetic reaction. The photosynthetic reaction can be conveniently summarized by the following general equation:

$$6\ CO_2 + 12\ H_2O + \text{Light Energy} \xrightarrow[\text{Enzymes}]{\text{Photosynthetic}} C_6H_{12}O_6 + 6\ H_2O + 6\ O_2$$

In this laboratory investigation we will examine several aspects associated with the process of photosynthesis. First, we will separate photosynthetic pigments using paper chromatography. We will determine the photosynthetic rate of *Elodea* by measuring the production of oxygen. Next, we will examine the location of starch within a *Coleus* leaf. Finally, we will examine the internal anatomy of the major photosynthetic appendage, the leaf.

Exercise I. Photosynthetic Pigments

In this exercise the four major plant pigments, chlorophyll a, chlorophyll b, carotene, and xanthophyll will be extracted from leaves. The pigments will be separated by paper chromatography. Paper chromatography is performed by absorbing a mixture of compounds on filter paper and a solvent is then allowed to move along the paper. Each compound will move with the solvent at a different rate. The rate that each compound is carried by the solvent depends upon the relative solubility of the compounds in the solvent. The most soluble compound will move fastest, etc. When the chromatography is terminated, the compounds will separate from one another and be located at different positions along the filter paper.

105

Objectives:

1. Identify and separate the pigments in spinach chloroplasts.
2. Become familiar with the procedure of paper chromatography.

Materials:

1. Frozen spinach leaves and acetone.
2. Beakers, cheesecloth, and pestle and mortar.
3. Chromatography paper (Whatman No. 1) and stapler.
4. Chromatography jars corks
5. Chromatography solvent (petroleum ether-acetone, 9:1).
6. Solvent waste jar.

Procedures:

1. The instructor will prepare a chloroplast pigment extract from spinach leaves by homogenising 2 leaves in 2ml acetone with a pestle and mortar.

2. Work in pairs. Obtain a sheet of chromatography paper, 12 × 1 cm in size. **Touch the paper as little as possible,** because oil from your fingers may interfere with development of the chromatogram.

3. Using a fine dropper, place a thin line of spinach extract along the width of the paper and about one-half inch from the bottom. Hold the paper in front of a fan or hair dryer in order to completely dry the paper. Repeat this step **three** times, placing the spinach extract along the same line each time in order to produce one dark line of spinach extract along the bottom of the paper.

4. After the three applications of spinach extract, **make sure the paper is completely dry.**

5. Pour the chromatography solvent into a chromatography jar to a depth of about 0.5 cm. Wait for the solvent to cease moving before carefully lowering the loaded paper into the jar. **DO NOT JAR THE CHROMATOGRAPHY JAR.** Quickly and carefully put the lid back on the jar. **CAUTION: DO NOT INHALE THE SOLVENT.**

6. The pigments move up the paper more slowly than the solvent. When the solvent "front" is about 2 cm from the top of the cylinder, remove and air-dry the chromatogram. Using a funnel, place the used chromatography solvent into the solvent waste jar.

7. Remove the staples from the chromatogram, and observe the colors and their relative positions on the paper. Ordinarily, the pigments appear on the paper as follows:

   a. orange **CAROTENES** (near the top).
   b. yellow-orange **XANTHOPHYLLS** (in the middle).
   c. blue-green **CHLOROPHYLL a** (just below the xanthophylls).
   d. yellow-green **CHLOROPHYLL b** (immediately below chlorophyll a).

Draw or attach chromatogram at the side of this page.

Exercise II. Carbon Dioxide Consumption and Oxygen Production During Photosynthesis

When examining the chemical equation for photosynthesis, one can see that carbon dioxide is consumed and oxygen is produced. In this experiment we will look at evidence that both events do occur. The aquatic plant *Elodea* will be used to study photosynthesis. When carbon dioxide is bubbled into water, the following chemical reaction occurs:

$$H_2O \quad + \quad CO_2 \quad \leftrightarrow \quad H_2CO_3 \quad \leftrightarrow \quad H^+ \quad + \quad HCO_3^-$$

| water | carbon dioxide | carbonic acid | hydrogen ion | bicarbonate ion |

When a pH indicator such as Phenol Red is added to the solution, the solution turns yellow indicating that it is acidic. As photosynthesis proceeds, the solution turns from yellow to red, indicating that carbon dioxide is being removed from the solution, causing the pH of the solution to shift to a neutral pH. If you observe the surface of the leaves of the *Elodea*, you can see small bubbles of oxygen collecting on the leaf surfaces, and even some bubbles rising to the surface.

Objectives:

1. Demonstrate the utilization of carbon dioxide in photosynthesis.

2. Demonstrate the production of oxygen as a result of photosynthesis.

Materials:

1. Photosynthesis apparatus: a 50 ml test tube, drinking straws, a ring stand and clamp to hold the test tube, a 500 ml beaker filled with water.

2. Dropper Bottles of Phenol Red (a pH indicator).

3. Sprigs of *Elodea*.

4. Razor blades.

5. Goose Neck Lamp with a 100 watt bulb.

Procedures:

1. Work in pairs.

2. Fill test tube half full of tap water and add 10 drops of Phenol Red. Using the drinking straw, gently blow into the Phenol Red solution. Carbon dioxide from your breath will make the solution acidic and the Phenol Red solution will turn yellow. **Stop** blowing as soon as the solution turns yellow.

3. Add a piece of *Elodea* (total length about 10 cm) to the test tube containing the Phenol Red (now yellow) solution.

4. Use the ring stand and test tube clamp to support the test tube and *Elodea* in the beaker filled with water. (This will keep the solution in the test tube from getting too hot).

5. Shine the flood lamp on the apparatus and observe every ten minutes over the next hour.

Questions:

1. How much carbon dioxide is found in the atmosphere? Oxygen?

2. What are the sources of carbon dioxide found in the atmosphere? Oxygen?

3. Why did the solution containing the *Elodea* turn red over time?

4. Why does carbon dioxide dissolved in the water have an effect on pH?

Exercise III. Chlorophyll and Photosynthetic Products

As indicated by the general equation for photosynthesis, one of the products of this process is glucose ($C_6H_{12}O_6$). Glucose, however, is quickly converted into a longer-term storage product, starch. Starch, consisting of long chains of glucose, is temporarily stored within chloroplasts. In this experiment, you will perform a test to determine the localization of starch in *Coleus* leaves.

Objectives:

1. Observe the localization of starch in *Coleus* leaves (green/white).

2. Determine the role of photosynthetic pigments in carbon fixation.

Materials:

1. *Coleus* plants that have been grown under high light conditions.

2. Hot plates.

3. Boiling water and ethyl alcohol baths and watch glasses.

4. Forceps.

5. Petri dishes.

6. Iodine (I_2KI) solution.

Procedure:

1. Obtain a variegated leaf from a *Coleus* plant. Draw the leaf in the space provided on next page, indicating the distribution of chlorophyll.

2. Using forceps, place the leaf in boiling water for about 30 seconds.

3. Transfer the leaf to a beaker of hot 95% ethyl alcohol, cover the beaker with a watch glass, and bring the alcohol to a gentle boil. **THE ALCOHOL IS HIGHLY FLAMMABLE, SO BE VERY CAREFUL.**

4. When the chlorophyll has been extracted, remove the now brittle leaf from the alcohol, placing it into some water in a Petri dish.

5. After about 30 to 60 seconds, pour off the water. The leaf will now be very soft, so care must be taken not to destroy it as the water is gently poured from the Petri dish.

6. Now flood the leaf with an iodine solution, and let it stand until portions of the leaf have darkened, then rinse the leaf gently with water.

7. Redraw the leaf, indicating the position of the purple-to-black reaction. Compare your second drawing with your first, and explain your results.

8. Discard the alcohol and iodine solution as indicated by your instructor.

Draw the leaf, indicating the distribution of starch in the leaf.

Coleus leaf
(distribution of chlorophyll):

Coleus leaf
(distribution of starch):

Exercise IV. The Leaf

The leaf is the principle lateral appendage of the stem. The primary function of the leaf is photosynthesis, wherein energy from sunlight is converted to chemical energy. Gaseous exchanges between the living leaf cells and the atmosphere are necessary to the process of photosynthesis. Important anatomical features which will influence the ability of a given leaf to carry on this function are: (1) total absorbing surface, both for gases and for sunlight, (2) permeability of epidermis to gases, (3) extent of intercellular spaces, and (4) nature and distribution of conducting tissue (xylem and phloem).

Objectives:

In this exercise, we will observe both external and internal features of leaves that influence the process of photosynthesis.

Materials:

1. Examples of different leaf types: monocots and dicots.

2. Prepared slides showing the internal anatomy of leaves.

Procedures:

1. Examine the leaves on display and note the various leaf shapes. In the leaves provided, note the **petiole, blade** or **lamina,** and the **main vein** or **midrib.**

2. Compare the grass leaf and the *Geranium* leaf. The main distinction between the leaves is in the characteristics of venation (arrangement of veins). Leaves of **monocots** (grasses, lilies, tulips, onions, etc.) have **parallel veins. Dicot** leaves will have **netted veins.**

3. Internal Structure: Observe the stained cross-section (Figure 13.1, p. 112). An overall view of this section reveals a prominent **midrib,** from which extends the **lamina.**

4. Examine the midrib first. It consists of two vascular tissue types: **xylem,** for conduction of water to the leaf, and **phloem,** for conduction of sugar away from the leaf. The xylem and phloem tissues are surrounded by a row of **bundle sheath cells.**

5. Next observe the lamina portion. The **mesophyll** is seen to be differentiated into two zones. Both zones have chloroplasts for photosynthesis. The upper, more compact layers are called the **palisade mesophyll.** The lower half of the mesophyll consists of an irregular network of cells, penetrated by air spaces. This layer is called the **spongy mesophyll.**

6. Search the epidermis for **stomata.** Are they on both the upper and lower epidermis? Observe the thick **cuticle** over the epidermis at the margin of the leaf.

7. **Epidermis.** Observe a piece of epidermis which has been stripped off the lower surface of a leaf. Note the irregular, jigsaw-puzzle form typical of epidermal cells. Ordinary epidermal cells lack chloroplasts. Now locate the **stomata.** Note the two **guard cells** which sit around the lens-shaped opening. (See page 112) Note that the guard cells do contain chloroplasts. Provide an explanation for why guard cells have chloroplasts while surrounding cells do not.

Make diagrams of the leaves provided.

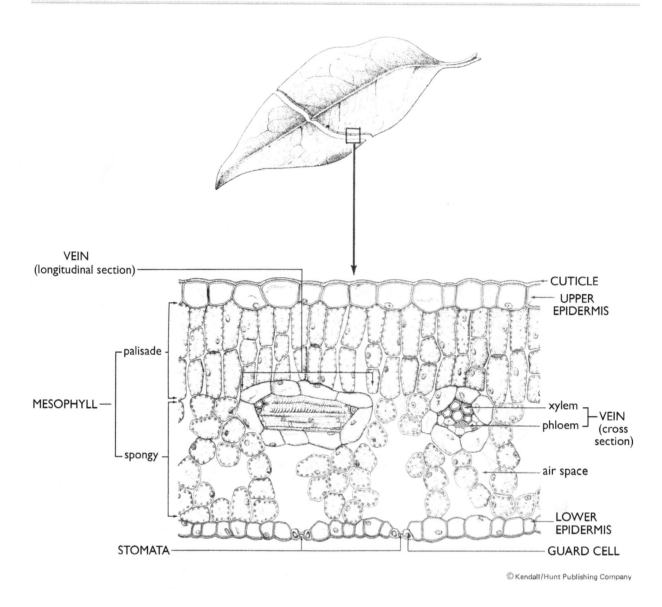

VEIN
(longitudinal section)

CUTICLE

UPPER
EPIDERMIS

palisade

MESOPHYLL

xylem

phloem

VEIN
(cross
section)

spongy

air space

LOWER
EPIDERMIS

STOMATA

GUARD CELL

© Kendall/Hunt Publishing Company

Figure 13.1: Cross-sectional View of a Typical Dicot Leaf

14

Cellular Respiration (Inquiry-Based)

Key terms to be mastered: Sugars (Glucose), Covalent Bond, Oxidation, Anaerobic Glycolysis, Pyruvate, Krebs (Citric Acid) Cycle, Electron Transport Chain, ATP, ADP, Inorganic Phosphate (Pi), Alcoholic and Lactic Acid Fermentation, Calcium Oxide, Cobalt Chloride, Hydrate, Carbonic Acid, Hydrogen Ion, Phenolpthalein, pH Indicator, Sodium Hydroxide

Objectives: The purpose of this chapter is to investigate respiration in humans and to compare respiration at rest and after exercise

Introduction

Being able to derive energy from the environment is another defining characteristic of life. When organic molecules, such as sugars, are oxidized by living organisms, **covalent bonds** are broken, and their stored energy is released. This energy is then used to make **adenosine triphosphate (ATP)** which can be used to do cellular work. This process (a type of biochemical or metabolic pathway) occurs in a number of steps, each being catalyzed by a specific enzyme. The entire process consists of three steps: 1) **glycolysis** 2) **the Krebs cycle** (citric acid cycle) 3) the **electron transport chain**. In the presence of oxygen, the three steps are completed in a process known as **aerobic respiration** which is represented by the following equation:

$$36\,ADP + 36 \searrow$$
$$6\,O_2 + C_6H_{12}O_6 \rightarrow 6\,CO_2 + 6\,H_2O$$
$$\searrow 36\,ATP$$

Glycolysis splits glucose into two **pyruvates** (pyruvic acids) and produces a net total of 2 ATPs. During glycolysis, electrons are also captured and stored for later use. This step occurs in the cytosol and requires no oxygen; it is therfore said to be anaerobic. The pyruvates are then oxidized to CO_2 gas in the mitochondrial matrix during the **Krebs cycle**, and electrons are also captured and stored. Finally, the stored electron energy from glycolysis and the Krebs cycle is used to drive the production of a large amount of ATP via the **electron transport chain** located on the inner mitochondrial membrane. Oxygen is required for this step to function.

The equation above shows that for each glucose ($C_6H_{12}O_6$) molecule oxidized, about 36 ATP molecules are formed. Energy derived from chemical reactions of the central equation above (large bold print) is used to drive the peripheral equation showing ATP production (smaller italic print). It is interesting to note that only 40% of the energy released from the oxidation or "burning" of glucose is used to form ATP; the remainder is lost as heat.

If oxygen is absent, glucose can still be split to yield 2 ATPs per glucose molecule. This process is known as **anerobic glycolysis**; examples include **alcoholic** and **lactic acid fermentation**. Byproducts of cellular respiration are CO_2 (carbon dioxide) and water. When produced by cells (especially muscle tissue!), CO_2 diffuses into the bloodstream, is taken to the lungs, and exhaled into the atmosphere during **expiration** (when we breathe out).

Activity 1: Investigating the Byproducts of Respiration: the Calcium Oxide Experiment

In the following experiment, you will add some calcium oxide to water and then blow into it. When you first add the calcium oxide to the water, the following will happen:

$$CaO + H_2O \rightarrow Ca(OH)_2$$

(Calcium oxide + water → Calcium hydroxide)

Then, when you blow into this solution, the following will happen:

$$Ca(OH)_2 + CO_2 \rightarrow CaCO_3 + H_2O$$

(Calcium hydroxide + Carobon dioxide → Calcium carbonate + Water)

Materials:

1. Calcium oxide

2. A straw

3. A test tube

Methods:

1. Work in groups of two.

2. Place 5 ml of distilled water in a test tube.

3. Add 0.1 g of calcium oxide to the water in the test tube.

4. Using your straw, slowly and carefully blow air into the beaker containing the calcium oxide solution

5. Record your observations in the section at the end of this chapter entitled *Questions for Thought and Discussion.*

Activity 2: Investigating the Byproducts of Respiration; the Cobalt Chloride Experiment

Cobalt chloride ($CoCl_2$) is a blue compound. When water is added to it, the cobalt chloride molecules will rearrange themselves to "make room" for the water. Water molecules will not bond covalently to the cobalt chloride but will instead surround the cobalt chloride forming what is known as a **hydrate** as follows:

$$CoCl_2 + 2\ H_2O \rightarrow CoCl_2 \cdot 2H_2O$$
(Cobalt chloride + Water → Cobalt dehydrate)

<u>Or</u>

$$CoCl_2 + 6\ H_2O \rightarrow CoCl_2 \cdot 6H_2O$$
(Cobalt chloride + Water → Cobalt hexahydrate)

Cobalt chloride hydrate comes in two forms, depending on the amount of water added. Cobalt dihydrate is surrounded by two water molecules and will appear purple. Cobalt chloride hexahydrate is surrounded by six water molecules and will appear pink. In this experiment, you will examine one of the byproducts of cellular respiration using a strip of paper containing cobalt chloride.

Materials:

1. One strip of cobalt chloride paper

Methods:

1. Work individually.

2. Blow onto the cobalt chloride paper (as if you were blowing out a candle).

3. Record your observations in the section at the end of this chapter entitled *Questions for Thought and Discussion.*

Activity 3: Human Cellular Respiration Experiment

Refer to the formula for cellular respiration in the introduction. A byproduct of respiration is Co_2 gas (carbon dioxide). Therefore, the presence of CO_2 in exhaled air should be an indicator of cellular respiration and proportional to the amount of O_2 (oxygen) consumed.

It is also known that CO_2, when mixed with water, will form H_2CO_3 (**carbonic acid**). Once formed, carbonic acid will dissociate into H^+ (**hydrogen ions**) and HCO_3^- (**bicarbonate ions**) as follows:

$$CO_2 + H_2O \rightarrow H_2CO_3 \rightarrow HCO_3^- + H^+$$
(Carbon dioxide + Water → Carbonic acid → Bicarbonate ion + Hydrogen ion)

Simply stated, carbon dioxide will make watery solutions acidic. The acidity of a solution can then be used as an indirect indicator for the presence of carbon dioxide. **Phenolphthalein** is a **pH indicator** which is clear in acidic solutions (pH < 7) and pink in neutral or alkaline solutions (pH ≥ 7). A pH neutral solution containing phenolphthalein should be pink. Exhaling into such a solution should lower the pH and make it clear (acidic). If a base, such as **sodium hydroxide** (NaOH) is added to the clear solution, the pH can be raised to neutral (pink) again.

So, after exhaling into a phenolphthalein solution and turning it clear, the number of <u>individual</u> drops of NaOH it takes to turn it <u>back</u> to pink can be used as an estimate of the amount of CO_2 produced and exhaled by a person. These facts can also be used to measure the effect of exercise on the rate of aerobic respiration.

Now, using the background information you have learned so far (from the introduction and this section), how would you assess the effect of different amounts of physical activity on human cellular respiration? Enter your response here:

Using the scientific method, propose a suitable hypothesis regarding cellular respiration in individuals with differing fitness levels? Enter your response here:

Propose a prediction here:

Propose an experimental means for testing your hypothesis here:

Materials:

1. Phenolphthalein

2. 0.4% NaOH (Sodium hydroxide)

3. dH_2O (Distilled water)

4. Graduated cylinders

5. Five 250 ml beakers (one control and four experimental beakers for each group of four students)

6. Transfer pipettes (to be used as droppers)

7. Drinking straws

8. Stop watch

Methods:

1. Work in groups of at most four; each student will be a subject for this experiment. Also, each student will assess his/her own fitness level by circling either "I" for Infrequent Exercise or "R" for Regular Exercise in the data tables at the end of this chapter. With the help of your instructor, all class data will be pooled and recorded in the data tables. (Do not volunteer if you have any health condition which prevents you from engaging in physical activity).

2. Creating the color control:

   a. Using a graduated cylinder, place 50 ml distilled water in a 250 ml beaker, add 4 drops of phenolphthalein, and gently swirl to mix.

   b. If the water turns pink, it can be assumed that there is little or no CO_2 in the solution; proceed to step 2d. If the water is clear, proceed to the next step (step 2c).

c. At this point, a clear phenolphthalein solution indicates the presence of some CO_2 which must be neutralized using NaOH. Neutralize the clear phenolphthalein solution by using a transfer pipette to slowly add 0.4% NaOH to the solution one drop at a time.

d. Swirl the solution to completely mix it after each drop of NaOH is added.

e. Stop adding the NaOH <u>as soon</u> as the solution develops any pink color which persists for more than one minute. Set this beaker aside; this is your color **control** for your entire group.

3. The experiment:

a. Simultaneously repeat steps 2a through 2e in a separate beaker for <u>each</u> member of your group. Be sure to obtain the same pink color as your control beaker. (To easily determine the color, hold your experimental and control beakers against a white background to compare them).

b. Before proceeding, make sure that all group members have been sitting and resting for at least 5 minutes.

c. Each group member should now steadily blow air through a straw into the pink solutions in their experimental beakers for 10 seconds. The pink color should fade or disappear after blowing.

d. Now, slowly add 0.4% NaOH (one drop at a time!) to the solutions in the beakers and swirl to thoroughly mix after each drop. Using the Goup Data Table at the end of this chapter, record the number of drops it takes to restore the original pink color. The original pink color can be determined by comparing the color of the experimental beakers with the control beaker after the addition of each drop of NaOH.

e. Next, have all group members exercise at a vigorous pace by either climbing stairs, running in place, or jogging down the corridors for 5 minutes.

f. Immediately repeat steps 3c and 3d (above). Record the number of drops needed to restore the original pink color.

g. Pooled class data will be used for your report.

Individual Group Data

(The purpose of this data table is to keep track of your group results while the experiment is in progress. This table is for convenience only, but it will not be used to generate your report; the table on the following page will be used for your final report. When you have completed this table, have only <u>one</u> student from your group remove this page and give it to the instructor. The instructor will then use this information to construct a pooled class data table)

Group Member Name (Use a Nickname if You Wish):	Fitness (Exercise) Level: I = Infrequent R = Regular (Circle ONE)	Number of Drops of NaOH Required	
1	I / R	After resting:	
		After exercise:	
2	I / R	After resting:	
		After exercise:	
3	I / R	After resting:	
		After exercise:	
4	I / R	After resting:	
		After exercise:	

Pooled Class Data

(Your Instructor will query each student and write the results on the board; transfer that data into the table below. Names will not be used; instead, your instructor will assign a "subject number" to every student. You will then transfer the information contained in data table (below) to your final report and generate a graph of the results.)

Subject Number (Anonymous):	Fitness (Exercise) Level: I = Infrequent R = Regular (Circle ONE)	Number of Drops of NaOH Required	
1	I / R	After resting:	
		After exercise:	
2	I / R	After resting:	
		After exercise:	
3	I / R	After resting:	
		After exercise:	
4	I / R	After resting:	
		After exercise:	
5	I / R	After resting:	
		After exercise:	
6	I / R	After resting:	
		After exercise:	
7	I / R	After resting:	
		After exercise:	
8	I / R	After resting:	
		After exercise:	

Subject Number (Anonymous):	Fitness (Exercise) Level: I = Infrequent R = Regular (Circle ONE)	Number of Drops of NaOH Required
9	I/R	After resting:
		After exercise:
10	I/R	After resting:
		After exercise:
11	I/R	After resting:
		After exercise:
12	I/R	After resting:
		After exercise:
13	I/R	After resting:
		After exercise:
14	I/R	After resting:
		After exercise:
15	I/R	After resting:
		After exercise:
16	I/R	After resting:
		After exercise:
17	I/R	After resting:
		After exercise:
18	I/R	After resting:
		After exercise:
19	I/R	After resting:
		After exercise:
20	I/R	After resting:
		After exercise:
21	I/R	After resting:
		After exercise:
22	I/R	After resting:
		After exercise:
23	I/R	After resting:
		After exercise:
24	I/R	After resting:
		After exercise:

Once you have completed the information in the tables above, complete/calculate the following:

1. Total number of subjects self-identifying as fitness level "I": _____

   a. Average number of drops of NaOH required for these subjects:

   1) At rest: _____

   2) After exercise: _____

2. Total number of subjects self-identifying as fitness level "R": _____

    a. Average number of drops of NaOH required for these subjects:

        1) At rest: _____

        2) After exercise: _____

3. You are now ready to write a detailed report of your experiment! Using all collected data, be sure to graph your results and explain your findings in your report. Your instructor will provide you with a more detailed explanation of the expected format and content of your report. Good luck!

Questions for Thought and Discussion

Regarding Cellular Respiration:

1. In what organelle does most aerobic respiration take place?

2. Which is the only step (out of the three) of cellular respiration that actually requires oxygen?

3. Can any ATP be made with out oxygen? Explain your answer.

4. Which step of cellular respiration creates most of the CO_2?

5. Why would a bodybuilder's resting rate of cellular respiration be much higher than that of an unconditioned person? (Hint: The answer lies in information contained within the last paragraph of the introduction.)

Regarding the Calcium Oxide and Cobalt Chloride Experiments:

1. What happened when you blew into the solution of calcium oxide? According the chemical equations in Activity 1, what product was formed in the test tube after you blew into it? What byproduct of cellular respiration (in your breath) contributed to your results?

2. What happened when you blew onto the strip of paper containing cobalt chloride? According to the chemical equations in Activity 2, what product was formed on the paper after you blew onto it? What byproduct of cellular respiration (in your breath) contributed to your results?

15

Cell Reproduction

Introduction

Mitosis and meiosis are two mechanisms by which the nuclei of cells divide. In **mitosis,** the parental cell divides producing **two daughter cells.** The daughter nuclei are **identical** to each other and to the parental nucleus in chromosome number and genetic makeup. The number of chromosomes per cell varies from one species to another. Human cells have 46 chromosomes **(23 homologous pairs). Homologous chromosomes** look alike, but carry different forms of the same genes. In higher plants and animals, having two sets of homologous chromosomes is called the **diploid** condition and is designated by the symbol **2N.**

Meiosis, in contrast, is a part of the **sexual life cycle** in which the daughter nuclei produced are found in cells that differentiate as male and female gametes: the **sperm** and **eggs.** Meiosis results in **four daughter cells** each containing **one-half** the chromosome number (**haploid** or **N**) and **different genetic composition.**

The Cell Cycle

The complex series of events that encompasses the life span of an actively dividing cell is called the **cell cycle** (Figure 15.1). After a cell is produced, it enters a period of growth and development termed **interphase** during which it grows to its maximum size and duplicates its chromosomes in preparation for cell division. During interphase the following events occur:

1. **G1** (gap 1) **phase**—Occurs after mitosis and cytokinesis, and is the primary growth phase of the cell.

2. **S** (synthesis) **phase**—DNA replication occurs. This is necessary so that each new cell will contain a complete copy of the genetic material.

3. **G2** (gap 2) **phase**—The second growth phase, in which preparations are made for mitosis. Mitochondria and other organelles replicate, chromosomes condense, and microtubules begin to assemble at a spindle.

4. **Cell division** occurs at the end of the cell's life cycle. **Nuclear division** is called **mitosis,** and is divided into 4 subphases: **Prophase, Metaphase, Anaphase,** and **Telophase.** Division of the cytoplasm into two daughter cells is called **cytokinesis.**

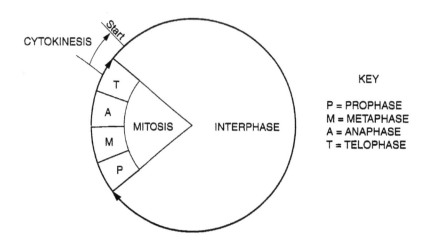

Figure 15.1:   The Cell Cycle

Mitosis in Plant Cells

The Alium (onion) root tip is an excellent model to use to study the events of mitosis. It contains an area called a **root meristem,** where there is active cell division. Within this region, you will be able to find every stage of mitosis.

Procedure

1. Obtain a prepared slide of a longitudinal section through an onion root tip. Search for the stages of mitosis, as outlined below. Draw each stage in the spaces provided.

2. **Interphase**

    A cell in interphase is "between divisions." The cytoplasm appears granular and a nuclear membrane is present. In these preserved and stained cells, the genetic material within the nucleus usually appears in the form of scattered CHROMATIN GRANULES. In the living state, fine filaments can be found which become thicker as cell division approaches. They are then known as CHROMOSOMES. Duplication of chromosomes occurs during interphase. The pairs of (sister) CHROMATIDS that result are not visible in the slide preparation you are studying. They remain joined together at a point called the CENTROMERE.

3. **Mitosis Stages in the Onion Root Meristem**

    Mitosis is a continuous process which is completed in a relatively short time. For convenience, the process is divided into stages.

    As you study the slide, it is not necessary to find the stages in order. However, it is important that they appear in order on your drawing page. If you happen to see a good telophase while looking for some other stage, stop, study, and draw it in the correct space on the drawing sheet provided. Use posters and models to help you identify the stages.

From *Biological Investigations*, Revised Printing by Gayne Bablanian. Copyright © 2002 by Gayne Bablanian. Used with permission of Kendall/Hunt Publishing Company.

a. Prophase

The nuclear membrane and nucleolus have dissolved but distinct chromosomes, each composed of paired chromatids, are now visible in the space formerly occupied by the nucleus. The onion has 16 chromosomes (now 32 sister chromatids) but counting them is impossible.

DRAW one cell and label: chromosome.

b. Metaphase

A fully-formed mitoic spindle now exists with fibers extending from one end of the cell to the other. These may not be visible. The point near each end of the cell where the spindle fibers converge is a spindle POLE (the term "pole" is a figure of speech like "north pole" or "south pole"). The sister chromatids are grouped in the center on an EQUATORIAL PLANE. SPINDLE FIBERS connect the CENTROMERES with the poles, however neither are visible with our microscopes.

DRAW a metaphase cell and label the region of the equatorial plane.

c. Anaphase

In this stage, the centromeres joining each pair of chromatids split and chromatids become chromosomes. The chromosomes are moved along spindle fibers toward opposite poles. Each chromosome bends at its centromere assuming the shape of a letter "V" as it is moved through the cytoplasm.

DRAW and label: chromosomes and poles.

d. Telophase

Having reached the poles, each group of chromosomes condenses into a mass in which the individual chromosomes are no longer visible.

Division of the cytoplasm (cytokinesis) begins in telophase. A **CELL PLATE** forms in the center of the cell at right angles to the spindle axis.

DRAW one cell and label: cell plate and chromosomes.

4. Cytokinesis: Dividing the Cytoplasm

Plant cells undergo cytokinesis by forming a cell plate across the equator of the cell. A cell plate is a collection of vesicles formed by the Golgi apparatus that contain cellulose and other cell wall components. The vesicles fuse to form plasma membranes and a cell wall that extends across the midplane of the cell. This divides the cytoplasm in half to form two daughter cells. Cytokinesis begins in telophase and is completed in interphase.

Find a pair of small cells in which the new cell wall has just been completed. These are new daughter cells.

DRAW a pair of daughter cells and label: new cell wall, old cell wall, and nucleus.

From *Experiencing Biology: A Laboratory Manual for Introductory Biology,* Seventh Edition/Revised Printing by GRCC-Biology 101 Staff, Biological Science Division. Copyright © 2002 by GRCC-Biology 101 Staff, Biological Science Division. Used with permission of Kendall/Hunt Publishing Company.

Mitosis in Animal Cells

Animal cells differ from plant cells in several ways, the most evident being that animal cells lack a cell wall. Unlike plant cells, animal cells have two pairs of **centrioles** that form the center of each pole during mitosis. The arrays of microtubules that completely surround each pole in an animal cell are called **asters.**

In animal cells, cytokinesis involves the formation of a **cleavage furrow.** A contractile ring of microfilaments pinches the cell in half to form two daughter cells. The contractile ring is similar to placing a string around a balloon and pulling it tightly to form two balloons.

You will be observing mitosis in cross sections through the blastula (a stage of development) of the whitefish.

Procedure

1. Obtain a prepared slide of whitefish blastula cross sections.

2. Under high power, identify the stages of mitosis, and draw each phase in the space provided.

3. Label the following structures in your drawings. Chromosomes, spindle fibers, asters, and cleavage furrow.

ANIMAL (whitefish blastula)

Interphase

Prophase

Metaphase

Anaphase

Telophase

DRAW one cell and label: cell membrane, nucleus, nucleolus, chromosomes and cytokinesis.

Estimating the Duration of the Phases of the Cell Cycle

Cells from the apical meristem of the onion root tip have a cell cycle that lasts approximately 20 hours. By counting the number of cells on your slide in each phase of the cell cycle and doing some calculations, you can estimate the number of hours that a cell spends in each phase of the cell cycle.

1. Return the Alium root tip slide to your microscope. Locate the apical meristem region of one of the root tips and study it at H.P. Count all of the cells in one H.P. field that are in interphase. Record the number in the table below and on the board.

2. Continue examining the same H.P. field of view. Count all of the cells in each of the 4 subphases of mitosis. Record these numbers in the table below and on the board.

3. If you have not counted at least 100 cells total, select another high power field at random and repeat the above steps. The more cells you count, the more accurate your estimates will be.

4. Record the total number of cells counted beneath the table.

5. Use the following equation to determine the duration (in hours) of each phase of the cell cycle.

$$\text{Duration} = \frac{\text{number of cells in a particular phase}}{\text{total number of cells counted}} \times 20 \text{ hours}$$

6. Record your results in the table below.

Phase of Cell Cycle	Number of Cells Counted in Each Phase	Duration in Hours
Interphase		
Prophase		
Metaphase		
Anaphase		
Telophase		

Total number of cells counted _____

Which phase of the cell cycle lasts the longest? _____

Which subphase of mitosis lasts the longest? _____

How do your results compare with published results?_____

_____

How could you improve the accuracy of your estimate?_____

_____

16

Patterns of Inheritance

Corn Genetics

The corn used in this part of the lab is called **genetic corn.** It is specially bred for genetics studies. Each kernel of the ear of corn represents an offspring produced by the union of an egg and a sperm. Thus, a single ear provides a large number of offspring resulting from a single cross. This provides the large sample size necessary for statistical accuracy.

This corn can be used to follow the inheritance of either or both of two different traits. One trait that can be studied is **kernel color.** There are two alleles for kernel color, *P* and *p.* Purple kernels (*P*) are dominant over yellow kernels (*p*). A second trait that can be followed is **kernel shape** or **texture.** There are two alleles for kernel shape, *S* and *s.* Smooth kernels (*S*) are dominant over wrinkled kernels (*s*). The genes for these two traits are located on separate chromosomes and are inherited independently of one another. Therefore, four possible combinations can occur in an individual kernel: purple smooth, purple wrinkled, yellow smooth, and yellow wrinkled. Locate all four phenotypes on the ear of genetic corn provided.

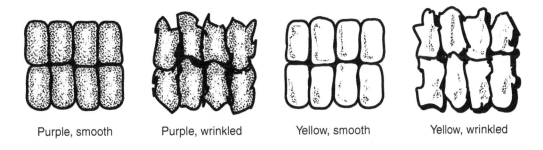

Purple, smooth   Purple, wrinkled   Yellow, smooth   Yellow, wrinkled

Figure 16.1: Genetic Corn

From *A Look at Life: Exploring the Unity of Organisms,* 4th Edition by Crowder, Durant, and Penrod. Copyright © 2000 by Carol S. Crowder, Mary A. Durant, and Shelley W. Penrod. Reprinted by permission of the authors.

Examining the Pedigree of the Corn

In order to analyze the inheritance of these traits, it is necessary to use experimental organisms that have a known genetic background. The corn used in this experiment actually represents the F_2 generation of a carefully manipulated cross. In the original **parental (P) generation,** one parent was homozygous (pure breeding) for purple, smooth kernels (***PPSS***). The other parent was homozygous (pure breeding) for yellow, wrinkled kernels (***ppss***). Since one parent will make all ***PS*** gametes and the other parent will make all ***ps*** gametes, all of the F_1 offspring will be ***PpSs*** and will have purple, smooth kernels as their phenotype. Finally, two of these F_1 offspring were crossed to produce the **F_2 generation**. In this activity, the phenotypic ratio of the F_2 offspring will be predicted. Then, data will be gathered to determined if the observed results match the expected ratios.

Predictions Section 1: Monohybrid Crosses

Inheritance of Kernel Color

1. Both of the F_1 parents were heterozygous ***Pp*** for kernel color (***Pp*** × ***Pp***). Determine the probability of each of the various genotypes appearing in the offspring.

    P(***PP***) = _____ %   P(***Pp***) = _____ %   P(***pp***) = _____ %

    Expected genotypic ratio = _____ ***PP*** : _____ ***Pp*** : _____ ***pp***

    Since purple is dominant to yellow, what is the probability of each phenotype?

    P(purple) = _____   P(yellow) = _____

    Expected phenotypic ratio = _____ purple : _____ yellow

2. Another method of predicting the results of a cross is the Punnett square. Fill in the Punnett square below and use it to predict the genotypic and phenotypic ratios.

    Key: ***P*** = purple
    　　　p = yellow

    Parents: ***Pp*** × ***Pp***

    Gametes = _____ and _____ (for **each** parent)

    Expected genotypic ratio = _____ ***PP*** : _____ ***Pp*** : _____ ***pp***

    Expected phenotypic ratio = _____ purple : _____ yellow

Inheritance of Kernel Shape

1. Both of the F_1 parents were heterozygous for kernel shape ($Ss \times Ss$). Since kernel shape is inherited in exactly the same manner as kernel color, what would be the probability of each of the genotypes from this cross?

    P(*SS*) = _____ P(*Ss*) = _____ P(*ss*) = _____

    Expected genotypic ratio = _____ *SS* : _____ *Ss* : _____ *ss*

    Since smooth is dominant to wrinkled, what is the probability of each phenotype?

    P(smooth) = _____ P(wrinkled) = _____

    Expected phenotypic ratio = _____ smooth : _____ wrinkled

2. Another method of predicting the results of this cross is the Punnett square. Fill in the Punnett square below and use it to predict the genotypic and phenotypic ratios.

    Key: *S* = smooth
         s = wrinkled

    Parents: *Ss* × *Ss*

    Gametes = _____ and _____ (for **each** parent)

    Expected genotypic ratio = _____ *SS* : _____ *Ss* : _____ *ss*

    Expected phenotypic ratio = _____ smooth : _____ wrinkled

Prediction Section 2: Dihybrid Cross

Inheritance of Kernel Color and Shape

In a dihybrid cross, the inheritance of two traits is followed simultaneously. In this exercise, both the inheritance of kernel color and kernel shape will be followed. As mentioned previously, the genes for these two traits are carried on separate chromosomes and are inherited independently of one another. As with the monohybrid crosses above, the laws of probability can be used to predict the outcome of a dihybrid cross. The F_1 generation was heterozygous for both kernel color and kernel shape (*PpSs*). Since each of the traits is inherited independently, the probability of inheriting any combination of the two traits is simply the product of the separate probabilities.

1. First, look back to the predictions made in the monohybrid crosses above and determine the probability of inheriting each of the following phenotypes. The first one is done for you.

    P(purple) = _3/4_   P(smooth) = _____

    P(yellow) = _____   P(wrinkled) = _____

2. Now, determine the probability of inheriting each of the following combinations of the two traits. Again, the first one is already done.

    P(purple and smooth) =   3/4 × 3/4   =   _9/16_

    P(purple and wrinkled) =   ___×___   =   _____

P(yellow and smooth) = _____×_____ = _____

P(yellow and wrinkled) = _____×_____ = _____

Expected phenotypic ratio = _____ : _____ : _____ : _____

3. As with the monohybrid cross, the Punnett square can also be used to predict the outcome of this cross. Complete the Punnett square below.

Key: **P** = purple **S** = smooth
      p = yellow **s** = wrinkled

Parents: **PpSs** × **PpSs**

Gametes: _____, _____, _____, & _____
        (same for each parent)

Expected phenotypic ratio = _____ : _____ : _____ : _____

Collecting Data

To assess the accuracy of the above predictions, data must be obtained. As stated above, there are four possible phenotypic combinations in the offspring: *purple smooth, purple wrinkled, yellow smooth, and yellow wrinkled.*

1. Using the ear of corn provided, count the number of kernels of each phenotype on 5 long, full rows. Record this information on Chart A of the data sheet. This group data will be used to analyze the predictions made for the monohybrid and dihybrid crosses.

2. Write the totals on the board so that class totals can be obtained. Record the class totals on Chart A of the data sheet. This class data will be used to analyze the predictions made for the monohybrid and dihybrid crosses.

Analyzing Data Section 1: Monohybrid Crosses

Inheritance of Kernel Color

Using the monohybrid corns provided (purple/yellow and smooth/wrinkled), count 5 long rows as before and record your results in Charts B and C.

1. To determine the observed ratio for the cross involving purple versus yellow kernel color, first calculate the number of purple and yellow kernels counted for both the group and the class. Referring to Chart A of the data sheet, add the number of purple smooth to the number of purple wrinkled. This gives the total number of purple kernels counted. Add the number of yellow smooth to the number of yellow wrinkled. This gives the total number of yellow kernels counted. Record this information on Chart B of the data sheet.

2. Calculate the **phenotypic ratio** for the group data by dividing each of the two numbers by the smaller number. *This will give a ratio of some number to 1.* Round off to **one decimal place** and record this information on Chart B of the data sheet. How close did the observed group ratio come to the expected ratio?

3. Now calculate the phenotypic ratio for the entire class using the class totals. Round off to one decimal place. Record this information on Chart B of the data sheet. How close did the observed class ratio come to the expected ratio?

Inheritance of Kernel Shape

1. To determine the observed ratio for the cross involving smooth versus wrinkled kernel shape, first determine the number of smooth and wrinkled kernels counted. Referring to Chart A of the data sheet, add the number of purple, smooth to the number of yellow, smooth. This gives the total number of **smooth** kernels counted. Add the number of purple, wrinkled to the number of yellow, wrinkled. This gives the total number of **wrinkled** kernels counted. Record this information on Chart C of the data sheet.

2. Next, calculate the **phenotypic ratio** for the group data by dividing each of the two numbers by the smaller number. Round off to **one decimal place.** *This will give a ratio of some number to 1.* Again, round off numbers to **one decimal place.** Record this information on Chart C of the data sheet. How close did the observed group ratio come to the expected ratio?

3. Now calculate the phenotypic ratio for the entire class using the class totals by dividing each of the two numbers by the smaller number. *This will give a ratio of some number to 1.* Again, round off numbers to **one decimal place.** Record this information on Chart C of the data sheet. How close did the observed class ratio come to the expected ratio?

Analyzing Data Section 2: Dihybrid Crosses

Inheritance of Kernel Color and Shape

1. Finally, to determine the observed ratio for the dihybrid cross involving both kernel color and shape, refer to Chart A of the data sheet and review the number of each of the four phenotypes that was counted. Use this information to determine the group and class ratios. To determine the ratio, divide each of the four numbers by the smallest number. *This will give a ratio of three numbers to 1.* Again, round off numbers to one decimal place.

2. Record this information on Chart D of the data sheet. How close did the observed class ratio come to the expected ratio?

Questions for Study and Review

1. Why would studies of inheritance involving garden peas or fruit flies be more statistically accurate than similar studies involving human families?

2. Which kernel color was dominant in the genetic corn? Which was recessive?

3. Which kernel shape was dominant in the genetic corn? Which was recessive?

4. If two corn plants that are heterozygous for both purple kernels and smooth seeds are crossed, what is the probability of an offspring having yellow kernels?

5. In the question above, what is the probability of an offspring having both purple kernels and wrinkled kernels?

6. Two plants heterozygous for purple kernels are crossed. Of the resulting 100 offspring, how many are expected to have yellow kernels?

7. In the above question, how much deviation from the expected would there be if 70 offspring have purple kernels and 30 have yellow kernels?

8. In the question above, what is the phenotypic ratio observed in the offspring?

9. In humans, blue eyes (*b*) are recessive to brown eyes (*B*). If two heterozygous brown-eyed people produce 3 children, what is the probability that all three will have blue eyes?

Observing Crosses Using Genetic Corn

Chart A: Genetic Corn Data

Phenotype	Group	Class
Purple Smooth		
Purple Wrinkled		
Yellow Smooth		
Yellow Wrinkled		

Chart B: Monohybrid Cross Kernel Color

Phenotype	Group	Class
Purple		
Yellow		
Ratio		

Chart C: Monohybrid Cross Kernel Shape

Phenotype	Group	Class
Smooth		
Wrinkled		
Ratio		

With both monohybrid crosses, what phenotypic ratio was expected? (Hint: refer to p. 124 & 125.)

_____

Which results were closer to this prediction, the group ratio or the class ratio? _____

Offer an explanation. _____

_____

Chart D: Dihybrid Cross (Kernel Color and Kernel Shape)

Group Ratio	Class Ratio

For the dihybrid cross, what phenotypic ratio was expected? (Hint: refer to p. 124 & 125.)

_____

Which results were closer to this prediction, the group ratio or the class ratio? _____

Offer an explanation. _____

_____

17

Human Inheritance

Introduction

Inheritance is more difficult to study in humans than in other organisms, such as corn or fruit flies, for several reasons. First of all, in contrast to genetic corn and fruit flies specially bred for genetics studies, humans have a very **limited knowledge of their genetic background.** While the pedigree of a strain of fruit flies may go back 100 generations, most humans are unable to trace their own ancestry back more than two or three generations. Second, humans have an extremely **long generation time.** In *Drosophila,* with a life cycle of only 10–14 days, a trait can be followed for several successive generations in a relatively short period of time so that patterns of inheritance can be determined. In humans, however, the average time between generations is 20 years. This makes it nearly impossible for a single researcher to follow the inheritance of a trait for more than one or two generations. Third, humans produce comparatively **few offspring.** Whereas a female fruit fly may produce as many as 1000 offspring, a human female produces just a few. Because of the small sample size, random deviation can cause significant distortion of the ratios in the offspring, making patterns of inheritance difficult to discern. Finally, in corn or fruit flies, experimental crosses can be performed between carefully chosen parents to reveal information about patterns of inheritance. However, for **ethical reasons,** experimental crosses are not performed on humans.

Until recently, the best way to determine how a particular trait was inherited was to study the **pedigree** of a family that showed the trait of interest. Examining an individual's "family tree" of ancestors and descendants can often provide important clues about how a particular trait is inherited. Not all traits, however, lend themselves readily to this type of analysis. Some traits are extremely rare so that families are difficult to find; or the trait may not be detectable without expensive biochemical testing. Additionally, not all families are large enough or know enough about their ancestry for a pedigree to be informative. More recently, with the advent of molecular techniques such as DNA sequencing that allow direct analysis of the DNA, much progress has been made in the investigation of human inheritance. As knowledge and understanding of human inheritance increase, advances are being made in the screening, diagnosis, and treatment of many serious human genetic disorders. Such advances may one day permit "gene replacement" therapy to correct genetic defects and cure genetic disorders. The study of human inheritance on the molecular level is certainly one of the most exciting new frontiers in biology today!

From *A Look at Life: Exploring the Unity of Organisms,* 4th Edition by Crowder, Durant, and Penrod. Copyright © 2000 by Carol S. Crowder, Mary A. Durant, and Shelley W. Penrod. Reprinted by permission of the authors.

Part I: Examining Some Common Inherited Traits in Humans

In the following exercise, some common human traits will be described. For each trait, the phenotype and possible genotype(s) will be determined. Next, class data will be collected for each trait so that the percentage of each phenotype can be calculated. Record this information on the chart in Part 1 of the data sheet. **Carry out percentage calculations to one decimal place.** For the curious student, more detailed information on these traits, as well as information about traits not covered in this lab, can be found in Victor A. McKusick's *Mendelian Inheritance in Man,* or online at the National Center for Biotechnology Information website (http://www.ncbi.nlm.nih.gov/).

1. Ability to Taste Sodium Benzoate

Sodium benzoate is a harmless chemical that is often added to foods as a preservative. The ability to taste sodium benzoate is genetically determined. Tasters have at least one copy of the dominant allele, *S,* while non-tasters are homozygous recessive *(ss).* Interestingly, tasters for sodium benzoate may differ as to whether it tastes sweet, salty, sour, or bitter!

First, obtain a small strip of *control taste paper* and place it on the tip, sides, and back of the tongue, but not in the middle, as there are no taste buds in the center of the tongue! Knowing what this tastes like will make it easier to compare with the test paper. Now, obtain a small strip of *sodium benzoate taste paper* and place it on the tip, sides and back of the tongue. Tasters may taste a sweet, sour, salty, or bitter taste.

2. Ability to Taste PTC

PTC (phenylthiocarbamide) is a harmless chemical that some individuals can taste and others cannot. Like sodium benzoate, the ability to taste PTC is genetically determined, but the two tastes are inherited separately. Tasters possess at least one copy of the dominant allele, *P,* while non-tasters are homozygous recessive *(pp).* In the United States, statistics show that about 71% of the Caucasian population and 90% of the African-American population are tasters. Some studies have indicated that individuals for whom PTC tastes bitter and sodium benzoate tastes salty tend to enjoy foods such as sauerkraut, buttermilk, spinach, and turnips.

Obtain a small strip of PTC taste paper and place it on the tongue, especially in the back. Tasters will detect a definite bitter taste.

3. Tongue Rolling

"Tongue rollers" have the ability to roll their tongue into a U-shape (Figure 17.1) while other individuals lack this ability. Tongue rolling is due to a dominant allele, *R,* while non-rollers are homozygous recessive *(rr).*

4. Widow's Peak

A hairline that comes to a distinct point in the middle of the forehead is called a "widow's peak" (Figure 17.2) and is due to a dominant allele, *W.* Individuals with a straight or even hairline are homozygous recessive *(ww).*

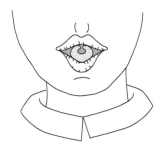

Figure 17.1: Tongue Rolling

Figure 17.2: Widow's Peak

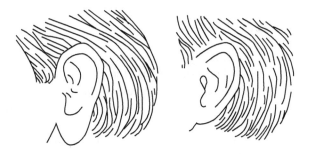

Figure 17.3: Free and Attached Earlobes

5. *Free (unattached) Earlobes*

Individuals that possess a dominant allele, *L,* have earlobes that hang free below the lowest point where the ear attaches to the head. Individuals that are homozygous recessive *(ll)* have attached earlobes in which the lowest point of the ear is where it attaches to the head. (See Figure 17.3.)

6. *Dimpled Cheeks*

Dimpled cheeks (*not* smile lines around the mouth) are due to a dominant allele, *G,* while lack of dimples is recessive *(gg).*

7. Cleft Chin

A cleft chin is due to the presence of a dominant allele, *V.* Individuals that are homozygous recessive *(vv),* lack this trait.

8. Hair Form

Hair form in humans is believed to represent an example of **incomplete dominance.** Individuals homozygous for the curly allele, *CC,* have curly hair. Individuals that are homozygous for the straight allele, *C'C',* have straight hair. Heterozygous individuals that are *CC'* have wavy hair, intermediate between curly and straight.

9. Red Hair Pigment

The main pigment found in hair that determines its color is melanin. Blondes produce small amounts of melanin while brunettes produce more. Red in the hair, however, is due to an iron-containing pigment called trichosiderin. Individuals with the dominant allele, *T,* produce an enzyme that breaks down trichosiderin. Individuals that are homozygous recessive *(tt),* do *not* produce the enzyme, resulting in the *presence* of trichosiderin.

  If trichosiderin is present in combination with little or no melanin, red hair results. If it is present along with melanin, then various shades of auburn result.

10. Iris Pigmentation

An individual's eye color is determined by the presence of a dominant allele, *J,* which causes the production of melanin in the outer layer of the iris (the pigmented portion of the eye surrounding the pupil). Individuals that are homozygous recessive, *jj,* fail to produce any melanin in the outer layer of the iris and, thus, have eyes that are blue to gray in color. Individuals that have the dominant allele, *J,* do produce melanin in the outer layer of the iris. Other genes determine exactly how much melanin is produced and how it is distributed in the iris. Additional genes code for other pigments (such as a yellow pigment called lipochrome) that blend with the melanin to produce other colors such as green or hazel.

11. Dominant Eye

When one views an object with both eyes open, each eye sees a slightly different view of the object. The brain sees only a single image, however, because one eye dominates the visual field. To determine which eye is the "dominant eye," obtain an index card with a hole cut in the center. Hold the card at arm's length and look at an object across the room *through* the hole with *both* eyes. Without moving the card, carefully close one eye, and then the other. Only one eye will actually be looking directly at the object and it "dominates" what is seen. The other eye is actually looking off to one side! A dominant allele, *E,* causes the right eye to be dominant, while individuals that are homozygous recessive *(ee),* are left eye dominant. This trait is *not* linked in any way to handedness.

12. Red/Green Colorblindness

The allele for normal color vision, X^C, is dominant over the recessive allele for red/green color-blindness, X^c. Because this is a **sex-linked** trait, a female must inherit two copies of the recessive allele in order to be red/green colorblind, while a male must only inherit a single copy of the recessive allele from his mother to show the trait. A female that is heterozygous for the trait has normal color vision. Red/green colorblindness often involves very subtle defects in color perception and many affected individuals are not aware of it until they are actually tested. It is at least 100 times more common for males to be red/green colorblind than females. To determine red/green colorblindness, use the Ichikawa colorblindness chart. Hold it at arm's length. All individuals, normal or red/green colorblind, read plate 1 as a "12." Normal individuals read plate 2 as an "8," while red/green colorblind individuals see a "3." Plate 3 reveals a "16" to a normal individual, while red/green colorblind individuals read the plate incorrectly or not at all.

13. Arm Folding

With arms folded across the chest, look to see which arm is on top. The dominant allele, **A,** results in the right arm on top, while the left arm on top is due to the homozygous recessive genotype, **aa.** Try folding the arms the other way and notice how uncoordinated it feels!

14. Hitchhiker's Thumb

Make a fist and extend the thumb as much as possible. If the last segment of the thumb is bent back at an angle of 60° or more from vertical, this is hitchhiker's thumb, and is a recessive trait **(nn).** Individuals with the dominant allele, **N,** lack this ability. (See Figure 17.4.)

15. Index Finger versus Ring Finger Length

Place both hands flat on a table, holding the fingers together. Compare the lengths of the index finger and the ring finger. Which is shorter? Which is longer? This trait is actually a **sex-influenced** trait. Because of hormonal differences, this trait is dominant in one sex and recessive in the other.

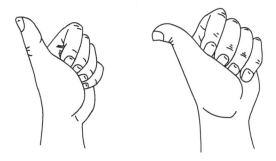

Figure 17.4: Hitchhiker's Thumb

	FF	*FF'*	*F'F'*
Males	Short Index	Short Index	Long Index
Females	Short Index	Long Index	Long Index

The allele *F* codes for a shorter index finger, while the allele *F'* codes for a longer index finger. In men, *F* is dominant, while in women, *F'* is dominant. Individuals that are homozygous *FF* have shorter index fingers, regardless of sex. Individuals that are homozygous *F'F'* have longer index fingers, regardless of sex. However, individuals that are heterozygous *FF'* will have shorter index fingers if they are males and longer index fingers if they are females.

16. *Hand Clasping*

With hands clasped together and fingers interlocked, look to see which thumb is on top. Like the trait above, some studies suggest that this is a **sex-influenced** trait. The allele **K** causes the right thumb to be on top, while the alternate allele, **K'**, results in the left thumb on top. In females, **K** is dominant to **K'**, while in males **K'** is dominant to **K**. (See Figure 17.5.) Now try clasping the hands the other way . . . how does it feel?

	KK	*KK'*	*K'K'*
Males	Right on top	Left on top	Left on top
Females	Right on top	Right on top	Left on top

Figure 17.5: Hand Clasping

17. Bent Little Finger

Place both hands flat on the table with fingers spread slightly apart. Carefully examine the last joint of the little finger. If it is bent inward, the dominant allele, **B,** is present. Individuals that are homozygous recessive *(bb),* have a straight last joint. (See Figure 17.6.)

18. Handedness

There is a genetic component to handedness. Individuals that are homozygous **HH** are strongly right-handed. Individuals who are homozygous **hh** are strongly left-handed. However, individuals that are **Hh** may be ambidextrous or they may be either left-handed or right-handed. Environment plays an important role in determining handedness in heterozygotes. In a society where being left-handed is sometimes discouraged by parents and teachers, children are often required to learn to do tasks with their right hand. Additionally, when learning a new skill, children will often mimic the handedness of the person teaching them. For this reason, they may write with their right hand but swing a baseball bat left-handed. For this exercise, consider handedness as the hand used for writing.

19. Mid-digital Hair

Carefully examine the upper surface of the middle section of each finger. Presence of hair on this section is due to a dominant allele, **M.** (See Figure 17.7.) Individuals who are homozygous recessive *(mm)* lack mid-digital hair. This gene may vary in expressivity. Some individuals may have hair on every digit, while others may have much less. Even one hair on one finger counts as a dominant phenotype!

20. Palmaris Longus Muscle

About 16% of males and 24% of females are missing an arm muscle called the palmaris longus in one or both arms. The presence or absence of this muscle can be determined by flexing a clenched fist and looking at the inner surface of the wrist for the presence of cord-like tendons. If three central definite tendons are detected, the muscle is present, but if only 2 tendons are present, the muscle is missing. **Not** having the muscle is due to the presence of a dominant allele, **U,** while individuals that are homozygous recessive *(uu)* have the muscle. (See Figure 17.8.)

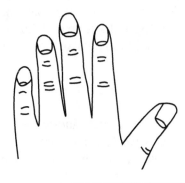

Figure 17.6: Bent Little Finger

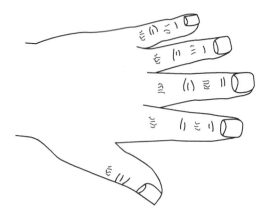

Figure 17.7: Mid-digital Hair

Figure 17.8: Palmaris Longus Muscle

21. *Polydactyly*

Having six digits on the hands and/or feet is termed polydactyly. This condition is caused by a dominant allele *(Y),* while the homozygous recessive genotype *(yy)* results in five digits. This gene does not show complete **penetrance.** About 13% of individuals that inherit the dominant allele fail to show the trait at all. This gene also has variable **expressivity.** Some individuals may express it strongly and have well-developed digits on both hands and feet, while others may only weakly express it and only have a partial vestigial extra digit on one hand or foot.

22. *Syndactyly*

Syndactyly refers to webbing between the fingers and/or toes. It is due to a dominant allele, **Z.** Individuals that are homozygous recessive *(zz)* have separated fingers and toes. The gene has variable expressivity. In some individuals, only two fingers may be joined, while in other individuals, all the fingers may be joined to form a hand that has a "lobster claw" appearance. The joining may actually involve the fusion of bones in addition to webbed skin.

Blood Type	A	B	AB	O
U.S. Caucasians	41%	10%	4%	45%
U.S. Blacks	26%	21%	4%	49%
U.S. Hispanics	40%	11%	5%	44%
Native Americans	7%	2%	0%	91%
Japanese	38%	22%	8%	32%
African Blacks	31%	29%	9%	31%
Chinese	30%	27%	8%	35%
Australian Aborigines	45%	2%	0%	53%

Rh+	Rh–
84%	16%
94%	6%
85%	15%
100%	0%
99%	1%
95%	5%
100%	0%
98%	2%

23. ABO Blood Type

The inheritance of the ABO blood type is an example of **multiple alleles.** There are three different alleles: I^A, I^B, and i. These alleles code for markers on the surface of the red blood cells called antigens. I^A codes for the A antigen, I^B codes for the B antigen, and i codes for no antigen. I^A and I^B are **co-dominant** to one another and both are dominant to i. Type A blood may be I^AI^A or I^Ai. Type B blood may be I^BI^B or I^Bi. Type AB blood is I^AI^B. Type O blood is ii. Above is a chart showing the frequencies of these 4 major blood types in different human populations around the world.

24. Rh Blood Type

An individual's Rh blood type is determined by the presence or absence of another marker on the red blood cell surface called the D antigen. A dominant allele. *D,* codes for the production of the D antigen resulting in Rh+ blood. Individuals that are homozygous recessive (*dd*) lack the D antigen on the RBC surface and are Rh–. This trait is referred to as the Rh blood type because the D antigen was first discovered in the blood of *Rhesus* monkeys before it was discovered in human blood. The Rh blood type is inherited separately from the ABO blood type. It is essential that both blood types be typed and matched for blood transfusions. The previous chart also shows the frequencies of Rh+ and Rh– blood in different human populations around the world.

Human Genetics Lab Data Sheet

Student Name: _____ Date: _____

Part 1: Some Common Inherited Traits in Humans

Trait	Phenotype	Genotype(s)	# Dominant in Class	% in Class
Sodium Benzoate Tasting (**S** or **s**)				
PTC Tasting (**P** or **p**)				
Tongue Rolling (**R** or **r**)				
Widow's Peak (**W** or **w**)				
Unattached Ear Lobes (**L** or **l**)				
Dimpled Cheeks (**G** or **g**)				
Cleft Chin (**V** or **v**)				
Hair Form (**C** or **c**)				
Red Hair (**T** or **t**)				
Iris Pigmentation (**J** or **j**)				
Dominant Eye (**E** or **e**)				
R/G Colorblindness (**X^C** or **X^c**)				
Arm Folding (**A** or **a**)				
Hitchhiker's Thumb (**N** or **n**)				
Index vs. Ring Finger (**F** or **F'**)				
Hand Clasping (**K** or **K'**)				
Bent Little Finger (**B** or **b**)				
Handedness (**H** or **h**)				
Mid-Digital Hair (**M** or **m**)				
Palmaris Longus Muscle (**U** or **u**)				
Polydactyly (**Y** or **y**)				
Syndactyly (**Z** or **z**)				
ABO Blood Type (**I^A**, **I^B**, or **i**)				
Rh Blood Type (**D** or **d**)				

From the traits studied in this lab, list examples for the following **patterns of inheritance:**

1. sex linked: _____   4. sex influenced: _____

2. incomplete dominance: _____   _____

3. multiple alleles: _____   5. co-dominance: _____

EXERCISE 17

Questions for Study and Review

1. If someone has a dominant trait such as Widow's Peak, how might it be determined whether they were homozygous or heterozygous?

2. Is there any correlation between the dominance/recessiveness of a trait and how common it is in the population?

3. Name 4 reasons why inheritance is more difficult to study in humans than in other organisms such as corn or fruit flies.

4. What is a karyotype? How many pairs of chromosomes are in a normal human karyotype? How many are autosomes? How many are sex chromosomes?

5. Which sex chromosomes are present in the cells of a male? Of a female?

6. Name three criteria that can be used to identify homologous chromosomes.

7. What is aneuploidy? What is trisomy? What is monosomy?

8. Briefly characterize individuals with the following conditions: Down Syndrome, Turner Syndrome, Klinefelter Syndrome, Triplo-X, and Jacobs Syndrome. Include the chromosome make-up of the individual as well as their sex.

9. Choose one of the traits from Part 1 of this exercise and create a pedigree for the inheritance of this trait in your immediate family. Use the blank pedigree form provided below. Three consecutive generations are shown. Shade in the circle to represent the recessive phenotype and leave it unshaded to represent the dominant phenotype. Modify each circle as shown to indicate male ($\male$) or female ($\female$). Add additional lines and circles to the pedigree as needed to indicate the relationships as they apply in your particular case.

FAMILY PEDIGREE

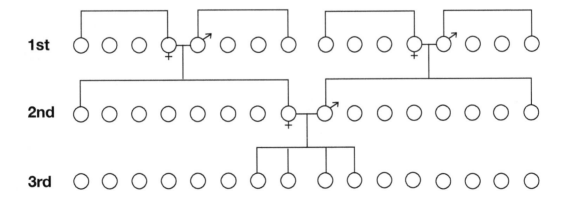

Part II: Making a Child

Each human being, with the exception of identical siblings, is genetically unique. We are going to examine some easily seen human traits or characteristics that give each person their uniqueness. We will then pair up and see what type of children we might have, if we were to "reproduce."

Use the following information to fill in the data table below. Notice that some of these traits exhibit complete dominance and some show incomplete dominance.

1. **Hair type:** Is your hair naturally curly (CC), wavy (Cc), or straight (cc)?

2. **Widow's peak:** Does your forehead hairline form a point (widow's peak) in the center? Present (WW, Ww) is dominant over absent (ww).

3. **Eyebrows:** Bushy (BB, Bb) is dominant over fine (bb).

4. **Color of eyebrows:** Are your eyebrows darker (HH), the same (Hh), or lighter (hh) than your hair?

5. **Eyelashes:** Long (LL, Ll) is dominant over short (ll).

6. **Eye color:** Non-blue (EE, Ee) is dominant over blue (ee). (However, note that we are over-simplifying for this lab; eye color is a quantitative trait, determined by at least 8 different loci).

7. **Freckles:** Present (FF, Ff) is dominant over absent (ff).

8. **Earlobes:** Free or unattached to the sides of the head (UU, Uu) is dominant over attached (uu).

9. **Dimples:** Present (DD, Dd) is dominant over absent (dd).

10. **Tongue rolling:** Ability to roll your tongue into a U-shape (RR, Rr) is dominant over non-rolling (rr).

If you have a choice, assume that you are heterozygous for the trait.

Table 17.1: Human Traits

Trait	Phenotype (describe)	Possible Genotypes
Hair type		
Widow's peak		
Eyebrows		
Color of eyebrows		
Eyelashes		
Eye color		
Freckles		
Earlobes		
Dimples		
Tongue-rolling		

Making a Child

1. Complete the following data sheet for these 10 traits as before (Table 17.1)

2. Conduct the "mating" process, using 11 sticks (one for each trait and one for the sex chromosomes, X or Y) as follows:

3. Put your initials on both sides of the stick, and for each trait write one allele on one side and the other allele on the other side. If you are dominant for a particular phenotype assume your genotype is heterozygous for this exercise.

4. Stick 11 should be labeled X or Y depending on your sex.

5. Each stick now represents a single trait that you will contribute to your child and each side of a stick one of your two homologous chromosomes.

6. You and your partner should now mix your 22 sticks randomly and drop them on the floor or on the table, all at the same time.

7. The side of the stick that turns up determines which allele contributes to the offspring

8. Record the genotype for each trait on Table 2 and draw a diagram of your child.

9. Remember that in meiosis chance will determine which allele will end up in which gamete. Meiosis also determines which gamete (sex cell) will deliver all of its chromosomes to the other gamete during fertilization of the offspring.

Table 2

Child

Parents' Names _____ and _____

Child's Name _____ Sex _____

Trait	Mother's Allele	Father's Allele	Child's Genotype	Child's Phenotype
Hair Type				
Widow's Peak				
Eyebrow				
Color of Eyebrows				
Eyelashes				
Eye Color				
Freckles				
Earlobes				
Dimples				
Tongue Rolling				

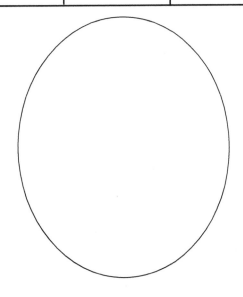

18

Introduction to Biotechnology

Key terms to be mastered: DNA, RNA, Nucleic Acid, Double Helix, Monomer, Nucleotide, Phosphate Group, Deoxyribose, Ribose, Nitrogenous Base, Adenine, Cytosine, Guanine, Thymine, Uracil, Base Pair, Hydrogen Bond, Genetically Modified Organism (GMO)

Objectives: The purpose of this chapter is to familiarize the student with the basic structure and function of nucleic acids. Basic concepts in biotechnology will be explored employing DNA extraction techniques.

Introduction

The presence of DNA (**Deoxyribonucleic acid**) in living organisms is one of the criteria used to determine whether or not something is alive. DNA represents the universal instructions of heredity used by all life on Earth for almost 4 billion years.

DNA was first isolated and extracted from cell nuclei by the Swiss scientist Frederick Miescher in 1869; he called this acid "nuclein". We now refer to these acids (DNA and **RNA**) as **nucleic acids**. Miescher and his contemporaries, however, were unsure of the function of nuclein. DNA's role as the molecule of heredity was not determined until after Frederick Griffith's 1928 mouse experiments. Griffith's results influenced scientists like Oswald Avery to conduct more rigorous experiments. In 1944, Avery and his colleagues proved that DNA was the molecule of heredity. Further work by James Watson and Francis Crick led to the discovery of the actual **double helix** structure of DNA in 1953. Watson and Crick were awarded the Nobel Prize for medicine in 1962.

It is now known that double stranded DNA and single stranded RNA are composed of smaller building block **monomers** called **nucleotides** (see Figure 18.1). Nucleotides are composed of a **phosphate group**, a pentose sugar (**deoxyribose** for DNA or **ribose** for RNA), and a **nitrogenous base**. There are five different bases: **Adenine** (A), **Cytosine** (C), **Guanine** (G), **Thymine** (T; DNA only!), and **Uracil** (U; RNA only!). The nucleotides in DNA are joined so that the phosphates and sugars make up the sides of the DNA "ladder", while two bases (or **base pairs**) are combined via **hydrogen bonds** to make the "rungs" of the DNA ladder. The two halves of the DNA molecule twist around each other to form two spirals or a **double helix**. Although the

DNA "alphabet" consists of a mere four letters (A,C,G, and T), these letters can be arranged in long sequences to create an almost infinite number of different genetic instructions.

The ability to extract and purify DNA is a very important initial step in biotechnological research. Once DNA has been isolated, scientist can then analyze it in many different ways. DNA fingerprinting can help determine paternity, catch criminals, or identify diseases. Scientist can manipulate the DNA of living things (recombination) to help mass produce hormones, create new medications, or produce **genetically modified organisms** for enhanced agricultural output.

In this lab, you will first examine the structure of DNA by assembling your own model of DNA using the kits provided. After you have familiarized yourself with the 3-D structure of DNA and its components, you will become a novice biotechnologist by actually extracting DNA from cells of the plant material provided!

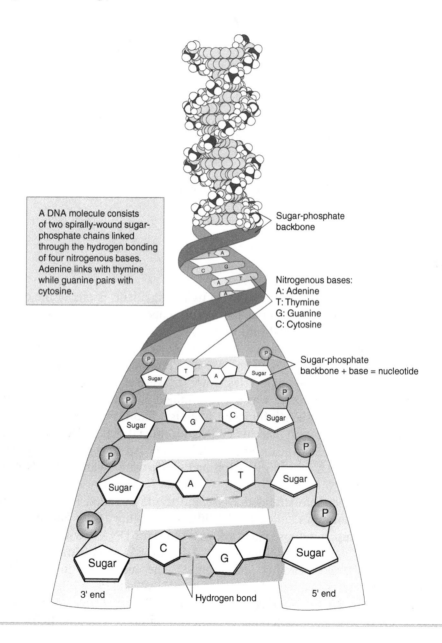

Figure 18.1

Activity 1: Investigating the Three Dimensional Structure of DNA

Materials:

1. DNA Kit (provided)
2. DNA Kit inventory sheet (to be signed by students assembling the kits)

Methods:

Note: Please be very careful with the DNA kits! Parts are very small and are easily lost

1. Work in groups of two.
2. Carefully open the DNA kit; count all the pieces, and sign the inventory sheet at the end of this chapter acknowledging that all parts are present and accounted for. **Inform your instructor immediately if pieces are missing!**
3. Your professor will have a completed DNA model at the front desk. Do not "cheat" by copying the model! Try your best to figure out how to assemble your own model first; if you get "stuck", refer to the completed instructor model for assistance. Now, using the DNA diagram (Figure 18.1) and your inventory sheet of DNA model parts as your primary guides, assemble your DNA molecule as follows:

   a. Construct your DNA model display stand:

      1) Connect the three <u>longer pale</u> green straws to the gray 4 pronged center.
      2) Connect the <u>long gray</u> straw to the gray 4 pronged center.
      3) Set aside the display stand for later use.

   b. Construct the "backbone" or two sides of your DNA molecule:

      1) Make two separate sides of your DNA molecule by connecting 5 two pronged red "phosphates" with 6 three pronged black "sugars" using 10 yellow "covalent bonds". The two completed sides should have a "zig-zag" appearance and should have black "sugars" at either end.

   c. Following the base pairing rules, attach bases to the free prongs of your black "sugars". Remember that you are essentially constructing a "ladder"!

   d. Connect both halves of your DNA molecule using the white two pronged "hydrogen bond" connectors.

   e. Display your finished DNA model on the previously constructed display stand by feeding the long gray straw through all the holes of the white "hydrogen bonds".

4. Have your instructor briefly inspect your finished model; be prepared to explain the model to your instructor!
5. Carefully disassemble your model, re-take inventory of all parts, and place all components in the appropriate bags. Seal the bags tightly! <u>Every student</u> should sign and turn in an inventory sheet. **Inform your instructor immediately if pieces are missing!**

Activity 2: Extracting DNA from Living Tissue

Part One: Cell Lysis
Materials (Cell Lysis):

1. Plastic sandwich bag with zipper style locking mechanism
2. A fresh piece of soft fruit, onion, or some berries
3. Sodium Chloride (Table salt)
4. Liquid Dishwashing soap
5. Distilled water
6. 250 ml beaker
7. Cheesecloth
8. Rubber band

Method (Cell Lysis):

1. Work in groups of two.
2. Pour the equivalent of one 10 cm test tube of distilled water into a locking sandwich bag.
3. Add a pinch of Sodium Chloride (table salt) to the water in the bag.
4. Add ONE drop dishwashing liquid to the bag.
5. Add a piece of soft fruit, onion, or some berries to the mixture in the bag (Suggested amount of fruit: ½ strawberry, or ¼ of a kiwi fruit, or a small wedge of onion, or a 2 cm section of banana, or 4 small berries such as raspberries, etc.)
6. Seal the sandwich bag.
7. Using your hands, gently squeeze and mash the contents of the bag for several minutes, and then set the sealed bag aside.
8. Obtain a 250 ml beaker from the supply area.
9. Cut a section of cheesecloth at least twice as large as the top opening of the beaker.
10. Cover the beaker with the cheesecloth and use the rubber band to hold the cheesecloth to the outside of the beaker.
11. Press down on the center of cheescloth covering the beaker to create a well or slight depression in the cheesecloth (See Figure 18.2).
12. Open the sandwich bag and carefully pour/strain the contents through the cheesecloth covering the beaker. This step requires patience and may take a few minutes.
13. Discard the cheescloth and plant solids, and save the strained liquid extract in the beaker for Part Two below.
14. Clean up your lab stations and discard all trash before proceeding to Part Two (below): SAVE the rubber bands and return them to the supply area but discard cheesecloth, fruit pieces, and sandwich bags. Salt and diswhasing liquid should be returned to the supply area.

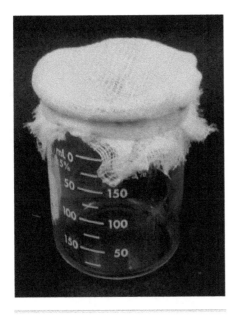

Figure 18.2   Figure 18.3

Part Two: DNA Extraction

Materials (DNA Extraction):

1. The 250 ml beaker containing strained fruit/vegetable (liquid only) from Part One

2. A clean 250 ml beaker

3. Two 10 cm test tubes

4. Ethanol (preferably cold)

5. Pipette pump (blue or green handle) with disposable serological tips

6. Disposable transfer pipette

7. A DNA spooler or a thin bamboo barbeque skewer

Method (DNA Extraction):

1. Using a pipette pump with a serological tip, fill a clean 10cm test tube ½ full with ethanlol. Place this partially filled test tube upright in an empty 250 ml beaker for later use.

2. Gently pour some of the juice extract (created in Part One) into a second clean 10cm test tube until the tube is about ½ full.

3. Place the test tube containing the colored juice extract in the 250 ml beaker next to the test tube containing the ethanol. This test tube should be resting inside the 250 ml beaker at an angle.

4. Using a clean transfer pipette, SLOWLY transfer **ethanol one drop at a time** to the test tube containing the juice extract. Do NOT allow the alcohol to splash or squirt into the fruit extract! Rather, the alcohol should slowly roll down the interior glass surface of the test tube. Your goal is to create a separate layer of ethanol on TOP of your colored fruit juice extract without mixing the alcohol and the extract (See Figure 18.3)

5. Soon, you should see bubbles coming out of the fruit juice extract. You should also see a white, stringy substance begin to separate from the colorered juice and rise to the surface into the alcohol. This is DNA!

6. Once a noticalbe amount of DNA has risen into the alcohol layer, use your DNA spooler or other long instrument (skewer, etc.) to examine the DNA; you may choose to pull the DNA out of the test tube to look at it. Observe its consistency.

7. Clean your lab stations, discard all trash, and return supplies to their proper locations.

8. You are now a biotechnologist!

DNA inventory sheet

Before and after assembling your DNA models, count and check off all part from the list below and enter the actual number of parts in the appropriate column.

DNA Display Stand Parts	Quantity	Quantity (Initial)	Quantity (Final)
1. Gray four pronged connector	(1)	_____	_____
2. Long pale green straw (5 cm)	(3)	_____	_____
3. Long gray straw	(1)	_____	_____

DNA Molecule Parts

1. Black three pronged "sugars"	(12)	_____	_____
2. Red two pronged "phosphates"	(10)	_____	_____
3. Yellow "covalent bonds" (2 cm)	(20)	_____	_____
4. Blue "thymine" base (2 cm)	(4)	_____	_____
5. Red "adenine" base (2 cm)	(4)	_____	_____
6. Green "cytosine" base (2 cm)	(4)	_____	_____
7. Gray "guanine" base (2 cm)	(4)	_____	_____
8. White two pronged "hydrogen bonds"	(6)	_____	_____

Student Signature: _____; Kit Number: _____

EXERCISE 18

Questions for Thought and Discussion

Regarding the Structure of DNA:

1. The basic building block of a nucleic acid (DNA or RNA) is called a _____ and consists of three parts. The three parts are:

   A.

   B.

   C.

2. Describe TWO basic differences between the structure of DNA and the structure of RNA:

   A.

   B.

3. Describe the basic difference between the function of DNA and RNA:

4. Name (write out) the names of the FOUR bases of DNA. List them in pairs according the base pair rules:

   A. Base Pair (DNA): _____ ——— _____

   B. Base Pair (DNA): _____ ——— _____

5. **Challenge Question!** The nitrogenous bases of DNA are held together by hydrogen bonds. Hydrogen bonds are considered to be some of the weaker chemical bonds. Why is hydrogen bonding important to the function of DNA? Why not make the bonds stronger (covalent)?

6. Notes:

Name: _____   Date: _____

Regarding DNA Extraction:

1. Define Lysis:

2. Explain the purpose of mashing the fruit; what part of the plants' cells is being affected by the mashing?

3. Explain the purpose of adding salt to the solution in the bag; specifically state *how* the salt helps to extract the contents of the fruit cells. Use correct terminology in your answer!

4. Explain the purpose of adding dish soap to the solution in the bag; specifically state *how* the soap helps to extract the contents of the fruit cells. Explain how the soap works differently from the salt; use correct terminology in your answer!

5. Explain the purpose adding ethanol to your fruit juice extract

6. Describe the appearance of the pure DNA you extracted. Were you able to see the actual double helix? Why or why not?

7. Notes:

19

DNA Fingerprinting

Background

In 1983, a geneticist at the University of Leicester in England named Alec Jeffreys began studying inherited genetic variation between individuals and the evolution of contemporary genes. He achieved this by looking at a genetic peculiarity known as the intron. Introns are DNA base sequences that do not specify part of the gene's final protein or RNA product.

Through his research, Jeffreys determined that some introns contain the same repeating DNA base pair sequences, but the number of repetitions varied from person to person. For example, in one person a particular base pair sequence (e.g., ATCGATCGATCGATCGATCG) may be repeated ten times, while in another person the same sequence may be repeated 25 times). Although the sequence of genes is fairly constant from person to person, introns are unique to each person, except in the case of identical twins.

Because of the uniqueness of each person's pattern, these variable DNA base pair sequences can be used to distinguish one person from another. They can be separated using electrophoresis, then matched with complementary radioactive probes to create an individual-specific DNA banding pattern. This technique is called "DNA fingerprinting."

> **Did You Know?**
>
> DNA paternity testing could not be used to determine which of two identical twins fathered a child since identical twins share the same DNA sequence.

DNA Fingerprinting

Of the three billion nucleotides in human DNA, more than 99% are identical among all individuals. The remaining 1% that is different, however, adds up to a significant amount of code variations between individuals, making each person's DNA profile as unique as a fingerprint. Due to the very large number of possible variations, no two people (with the exception of identical twins) have the same DNA sequence.

For every 1,000 nucleotides inherited, there is one site of variation, or polymorphism. These DNA polymorphisms change the length of the DNA fragments produced by the digestion with restriction enzymes. The exact number and size of fragments produced by a specific restriction enzyme digestion varies from person to person. The resulting fragments, called Restriction Fragment Length Polymorphisms (RFLPs), can be separated and their size determined by electrophoresis.

Most of the DNA in a chromosome is not used to code for genes. It is uncertain what, if any, use this "unused" DNA may have. Because these regions are not essential to an organism's development, it is more likely that changes will be found in these nonessential regions. These regions contain nucleotide sequences (e.g., GTCAGTCAGTCAGTCA) that repeat from 20 to 100 times. These restriction enzymes that flank these repeating sequences cut the DNA strand creating RFLPs.

The differences in the fragments can be quantified to create a "DNA fingerprint." Distinct RFLP patterns can be used to trace the inheritance of chromosomal regions with genetic disorders or to identify the origin of a blood sample in a criminal investigation. Scientists have identified more than 3,000 RFLPs in the human genome, many of which are highly variable among individuals. It is this large number of variable yet identifiable factors that allow scientists to identify individuals by the number and size of their various RFLPs.

This technique is being used more and more frequently in legal matters. DNA fingerprints can positively exclude someone but only establish a probability to include someone. Using DNA fingerprinting, the identity of a person who has committed a violent crime can be determined from minute quantities of DNA left at the scene of the crime in the form of blood, semen, hair, or saliva. The DNA fingerprint matched to a suspect can be accurate to within one in 10 billion people, which is about twice the total population of the world. Certain limitations in the technique prevent two samples from being identified as a "perfect match", yet it is possible to measure the statistical probability of two samples coming from the same individual based on the number of known RFLPs that exist in a given population.

DNA fingerprinting has many other applications. Since half of a person's genome comes from each parent, DNA fingerprinting can be used to determine familial relationships. It has a much higher certainty than a blood test when used to determine fatherhood in a paternity suit. DNA fingerprinting can be used to track hereditary diseases passed down family lines and can be used to find the closest possible matches for organ transplants. It can also be used to ascertain the level of inbreeding of endangered animals, aiding in the development of breeding programs to increase animals' genetic health and diversity.

Did You Know?

DNA paternity testing can establish a probability of over 99% that a man has fathered a child. The only way to establish 100% probability would be to perform a DNA test on every male on Earth.

DNA Fragment Length Determination

Under a given set of electrophoretic conditions such as pH, voltage, time, gel type, concentration, etc., the electrophoretic mobility of a DNA fragment molecule is standard. The length of a given DNA fragment can be determined by comparing its electrophoretic mobility on an agarose gel with that of a DNA marker sample of known length. The smaller the DNA fragment, the faster it will move down the gel during electrophoresis.

Using a technique called Southern blotting, the separated fragments are transferred to nitrocellulose paper, labeled with a radioactive probe, and developed against X-ray film. The probe, which is coded to bind to specific RFLPs being tested, will develop the film. The greater the concentration of DNA in that particular band, the darker the band will be. The resulting image, called an autoradiogram, shows a series of dark and light bands. This pattern is the DNA fingerprint of the tested individual. Comparing the distances between the bands in different samples determines the similarities between the samples.

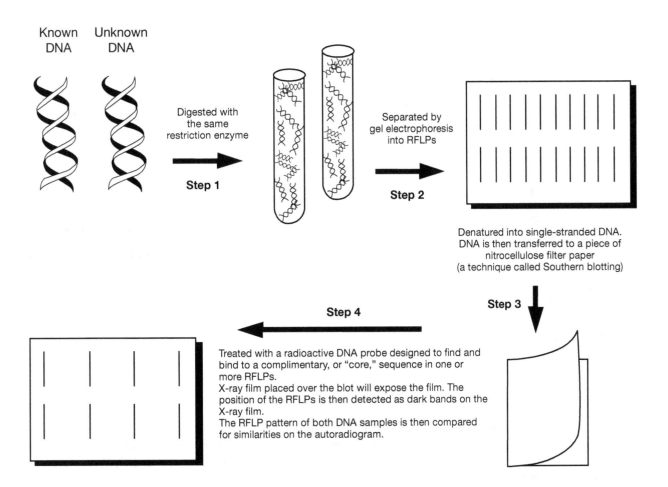

Prior to the advent of DNA fingerprinting technology, paternity testing was based on blood type, since the blood type of an offspring is based on the blood types of the parents. But blood types can only be used to exclude possible biological parents rather than to actually prove familial relationships with a high degree of certainty. DNA fingerprinting is used routinely to determine familial relationships. Paternity determination uses the principles of Mendelian genetics to infer the identity of the true biological father of an offspring, since typically the mother is known. Because half of a person's genome comes from each parent, any gene found in an offspring, not inherited from its mother, must have come from its father. The exact numbers of probes needed to resolve any particular case depends on gene frequencies and local breeding history. On the average, a typical paternity case will require four or five DNA probes for a decisive resolution.

Did You Know?

Unlike regular fingerprints, which can be altered and only occur on fingers, DNA fingerprints are the same for every cell of a person, whether it be skin, hair, bone, blood, etc.

Materials

Materials Needed Per Group

1   Flash Gel Electrophoresis System
    Power Supply
    Micropipet with tips
    Gloves

Shared Materials

DNA Samples:
  Mother
  Child
  Alleged father #1
  Alleged father #2
UV transilluminator (36 W 9951)
UV goggles (36 W 9951)
Biohazard bag (36 W 9951)

Did You Know?

Agarose is a highly purified form of agar, which is a material extracted from seaweed.

Scenario

Two men claim to be the biological father of a child. The courts have obtained DNA samples from the mother, child, and the two alleged fathers. As the director of a genetic testing laboratory, you will analyze the samples using the DNA fingerprinting technique to help settle the paternity dispute.

DNA analysis using the FlashGel® System

Important Safety Information and Quick Start Guide

The following symbols alert the user to important operational, maintenance, and/or warranty requirements, or possible hazards exposure.

CAUTION: Hazardous Voltage

Contact may cause death or serious injury. Caution should be exercised in the operation of this system as it can develop sufficient voltage and current to produce a lethal shock. To avoid any risk of injury, the system should only be operated by properly trained personnel and always in accordance with the instructions provided with The FlashGel® Dock, or found at www.lonza.com/protocols.

Prior to turning on the DC power source, ensure tshat **the black lead is connected to the negative terminal and the red lead is connected to the positive terminal**. Do not touch the FlashGel® Dock or Cassette while the high voltage supply is turned on. Do not add or recover samples to the FlashGel® Cassette while the high voltage leads are connected to the power supply. Mask to block light from the second tier of wells when using double-tier or FlashGel™ Recovery cassette.

Failure to adhere to the instructions could result in personal and/or laboratory hazards, as well as invalidate any warranty. Always turn off the DC power source prior to removing cassettes from the dock. For maximum safety, always operate this system in an isolated, low traffic area, not accessible to unauthorized personnel.

Never operate damaged equipment.

Precautions

The dock utilizes a visible light transilluminator to view fragments. It is safe to view cassettes on the lighted dock without UV light protection. Turn on the light only after the cassette is in place. Do not stare directly into the light.

Wear gloves, lab coat and safety glasses when handling FlashGel® Cassettes. The gel and buffer in FlashGel® Cassettes contain a proprietary DNA stain that is a potential mutagen. Follow state and local guidelines for handling and disposal of these materials.

Complete system instructions

Complete system instructions, including additional safety and warranty information are provided with The FlashGel® Dock. These can also be obtained on our website www.lonza.com/protocols, or by contacting Scientific Support. US: (800) 521-0390; Please see our website www.flashgel.com

Operating conditions for FlashGel® Dock

Environmental Conditions

Operating Conditions: Temperature: 15°C-35°C; Humidity: 15%-85% relative humidity, non-condensing

Storage and Shipping Conditions:

Temperature: 2°C-60°C

Humidity: 15%-85% relative humidity, non-condensing

Cleaning and Disposal - To avoid accidental exposure to high voltage, do not clean dock while connected to the high voltage power supply. Clean the FlashGel® Dock exterior with a cloth moistened with water or mild detergent. Do not immerse! The stain in FlashGel® Cassettes is a potential mutagen. Follow country, state and local guidelines for disposal of hazardous materials.

FlashGel® System Quick Start Guide

Refer to detailed instructions for sample preparation and run conditions (www.lonza.com/protocols).

- Cassette types are optimized for DNA, RNA or recovery.

- Do not exceed load volume of 5 μl for 12 + 1 and 16 + 1 cassettes and 12 μl for 8 + 1 cassettes.

- Optimal per band sample concentration is 1/5 that required for an ethidium bromide gel.

DNA Analysis with FlashGel® DNA Cassettes

1. Remove white seals from cassettes. Do not remove the clear vent seals.

2. Flood wells with distilled or deionized water. Blot away excess fluid by tilting the cassette to move excess fluid to the edge and blot off with a lint free wipe. Do not blot wells directly.

3. Load the required amount (preferably 10ul each of the diluted DNA sample, diluted 1:10) into the corresponding lane of the gel. Do not pierce the bottom of the wells with the micropipet tip. Do not overload.

4. If using double-tier cassettes, insert the FlashGel® Mask under the second tier.

   Mother – Lane #1; Child – Lane #3; Alleged Father 1- Lane #5; Alleged Father 2- Lane #7.

5. Insert cassette in to dock.

6. Plug in and turn on dock light.

7. Set high voltage power supply to 275 V, plug in cables and turn on power supply.

8. Run until desired separation is reached, then turn off high voltage power supply and disconnect the cables.

8. Photograph using The FlashGel® Camera or other documentation system.

   P.S. More than one group can run their samples on the same gel. And further, the gels can be used at least two times.

> **Did You Know?**
>
> If all of the DNA molecules found in a single human cell were lined up end-to-end, they would reach a distance of about two meters. Yet, all of these molecules are packed into a nucleus 10 millionths of a meter in diameter.

DNA Recovery with FlashGel® Recovery Cassettes

1. Follow steps 1 through 3 above.

2. Load samples to be recovered in the upper tier of sample wells.

3. Run until just prior to desired sample reaching the recovery wells (2nd tier), then stop the run and disconnect the high voltage cables.

4. Blot excess buffer from the recovery well(s) and add 20 µl of FlashGel® Recovery Buffer.

5. Remove the FlashGel® Mask, reconnect the cables and restart the power. Use the FlashGel® Visualization Glasses to observe band migration.

> **Did You Know?**
>
> The fluorescent dye ethidium bromide intercalates between the base pairs of DNA. UV light at 260 nm is absorbed by the DNA and transmitted to the EtBr. UV light at 300 nm and 360 nm is absorbed by the dye itself. Ethidium Bromide fluoresces by emitting a wavelength of 590 nm, which is in the red-orange range of the visible spectrum.

6. When the band of interest has migrated to the center of the recovery well, turn of the power supply and disconnect the cables.

7. Use a pipette to carefully remove the buffer containing the DNA. Recovery of high DNA loads may require repeating the process of loading recovery buffer and running the band further in to the well to maximize recovered volume.

RNA Analysis with FlashGel® RNA Cassettes

1. Follow the procedure for DNA analysis. RNA bands will be visible only for the first 3 to 4 minutes of the run.

2. Stop the run after 8 minutes and hold. The bands will be visible again after > 10 minutes, depending upon RNA load intensity.

Some components of the FlashGel® System are sold under licensing agreements. The nucleic acid stain in this product is manufactured and sold under license from Molecular Probes, Inc., and the FlashGel® . Cassette is sold under license from Invitrogen IP Holdings, Inc., and is for use only in research applications or quality control, and is covered by pending and issued patents. The FlashGel® Dock technology contains Clare Chemical Research, Inc. Dark Reader® transilluminator technology and is covered under US Patents 6,198,107; 6,512,236; and 6,914,250. The electrophoresis technology is licensed from Temple University and is covered under US Patent 6,905,585.

Dark Reader is a registered trademark of Clare Chemical Research, Inc.

All other trademarks herein are marks of the Lonza Group or its affiliates.

Analysis

1. Sketch your gel in the space below. Be sure to include all visible bands of DNA. Examine your gel carefully. Smaller bands of DNA may be harder to see.

Electrophoresis Gel

Mother	Alleged Father 1	Child	Alleged Father 2
▭	▭	▭	▭

2. Based on the electrophoresis results, determine which of the two alleged fathers is the actual father of the child. Explain how you came to this conclusion.

Name: _____   Date: _____

Assessment

1. Below is an illustration of an autoradiogram, which is typically used in paternity disputes. Based on your knowledge of DNA fingerprinting and the fact that half a person's genome comes from each parent, is the alleged father actually the biological father of the child? Explain your answer.

Mother	Child	Alleged Father	Child and Alleged Father

2. Based on the autoradiogram below, is the alleged father the biological father of the child?

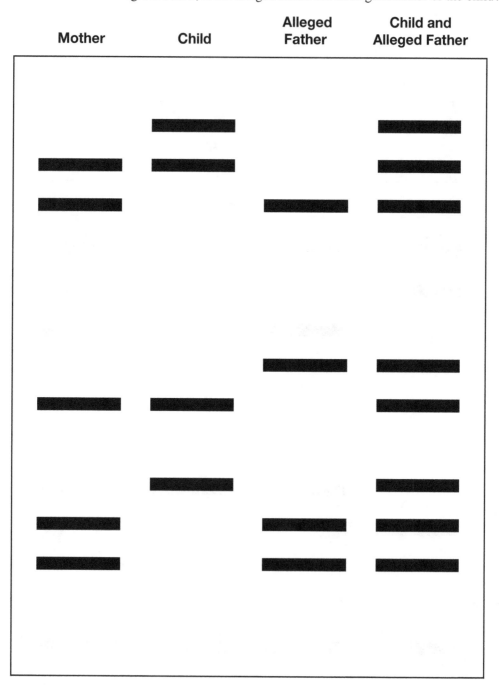

3. Three women, each of whom have one child, claim that Mr. X is the father of their children. Based on the banding patterns in the gel below, which of the children may or may not be Mr. X's?

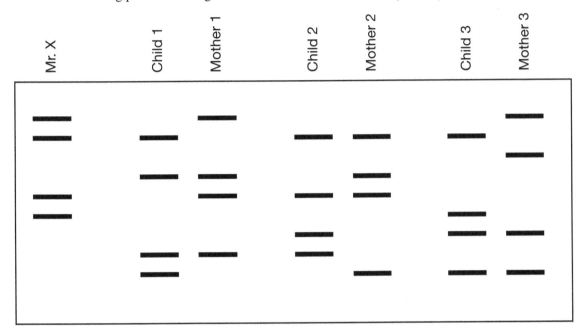

4. The restriction enzymes *Eco*R I recognizes GAATTC, and *Hind* III recognizes AAGCTT. A student adds *Eco*R I to a linear DNA sample. To another quantity of the same DNA, she adds *Hind* III. In a third tube she adds both enzymes. She runs a gel and the following occurs:

Hind III	**EcoR I**	**Double**
————		
	————	————
	————	
		————
————		————

The student, after viewing the gel, draws the following map of the linear DNA

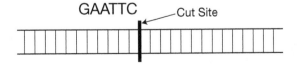

GAATTC   Cut Site

a. Explain why she placed the *Eco*R I restriction enzyme site as she did.

After a little more thought, she added another sequence to the DNA:

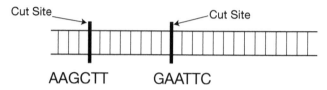

b. Explain how she was able to add Hind III restriction enzyme site in this position based on the results of her gel.

5. Different restriction enzymes are isolated from different types of bacteria. What advantage do you think bacteria gain by having restriction enzymes?

6. Predict what would happen if you place your gel in the electrophoresis chamber with the wells containing the DNA next to the red electrode instead of the black.

7. If you have a restriction enzyme that cuts a piece of linear DNA at two recognition sites, how many DNA fragments would you see on a gel?

8. Below is a list of the components involved in agarose gel electrophoresis. Briefly describe the purpose of each component.

Agarose gel—

TBE buffer—

Electrophoresis chamber—

Power supply—

DNA samples—

DNA stain—

9. Electrophoresis is one method of separating molecules. Paper chromatography is another method of separating molecules. Create a Venn diagram showing at least two similarities and two differences between these two methods.

10. You have used electrophoresis to examine paternal relationships and identify the most probable DNA match between two alleged fathers. Research another use for electrophoresis and briefly describe the benefits of using this technology.

20

Natural Selection

Key terms to be mastered: Evolution, Coevolution, Evolutionary "Arms Race", Population, Mutation, Natural Selection, Genetic Drift, Gene Flow, Predator, Prey, Polymorphic, Genetic Variation, Fitness, Allele

Objectives: The purpose of this chapter is to examine natural selection as a force of evolution. Natural selection will be simulated through a predator/prey exercise.

Introduction

In addition to the presence of DNA, the capacity for evolution is another defining characteristic of all life. All living things on Earth have ultimately evolved from simple aquatic prokaryotic organisms. **Evolution** is defined as genetic change occurring in a **population** over time. Individuals do not evolve; only populations evolve. **Mutations** provide the raw material or "fuel" for evolution by creating **genetic variation**; the "machinery" or mechanisms for evolution include **natural selection, genetic drift**, and **gene flow**.

Differences in reproductive success define natural selection. The ability to both survive and reproduce is known as **fitness**. Individuals with genetic traits (**alleles**) favored in a particular environment will have a higher fitness and be more likely to produce higher numbers of offspring. Individuals who are not as well adapted to their environments will experience lower reproductive success (lower fitness) and are likely to be selected against or "weeded out" by natural selection. Population **alleles** with a lower fitness will decline with each passing generation. Natural selection may eventually eliminate certain alleles from populations if those alleles display a relatively low fitness. It must be stressed that evolution and natural selection are never anticipatory; they do not "plan ahead". Natural selection is based on current interactions between organisms and their environments. For example, an animal which is favored by natural slection today may find itself at a disadvantage if the environment changes in the future.

Natural selection can be seen in **predator/prey** interactions. Consider the case of the garter snake and the newt (a small amphibian). A particular newt species is frequently eaten by the snake. This newt species has evolved a very strong poison in its skin as a defense. The snake, however, has evolved resistance to the newt's poison over time. So, toxic newts and resistant snakes are continuously favored by natural selection. Less poisonous newts are eaten by snakes, and the less resistant snakes die due to the newts' poison. The evolutionary processes of these

183

two animals are linked by their interactions. The evolution of one species in response to the evolution of the other is defined as **coevolution**. Coevolution is very common in predator/prey interactions and can also lead to **evolutionary "arms races"**, as was seen in the snake/newt example.

In this lab, you will explore natural selection through a simulated predator/prey activity. You will be a seed predator who eats beans. There will be variations with respect to the structures of your feeding appendages (your "beaks") in your group of predators. There will also be variation within the seed (prey) population with respect to seed color. The feeding environments for the predators may also vary with respect to background color and texture. After all predators have finished feeding, calculations will be performed to determine the fitness of both the predators (you!) and the prey (the beans). Based on your fitness calculations, you will be able to determine whether or not the populations of certain predators and prey will either increase or decrease in the future.

Activity: Predator Feeding and Fitness Calculations

Materials:

1. A selected feeding habitat (1 m² rectangular trays with textured bottoms of varying colors: black, tan, or white)

2. 300 prey items consisting of equal numbers of different kinds of dried beans (100 black beans, 100 tan beans, and 100 white beans)

3. A simulated **polymorphic** bird (predator) population consisting of students with various types of feeding appendages or "beaks" (plastic tweezers, standard (std.) clothespins, heavy duty (h.d.) clothespins)

4. A crop ("stomach") consisting of a small disposable cup

5. A glass jar or other container for the 300 beans

6. A stopwatch

7. A calculator (provided by the student)

Methods:

1. Work in groups of three or four.

2. Count out 300 beans: 100 black, 100 tan, and 100 white beans (this should be a group effort!)

3. Place all beans in a glass jar or other container, and carefully shake the container to mix the beans. Be careful not to spill or drop beans!

4. Select your "beak". Note: Each student or "bird" in a group should have a *different* type of "beak"! If a group consists of four individuals, no more than two individuals may have the same type of "beak".

5. Each "bird" (student) should have a crop (paper cup) to store captured beans.

6. Scatter the 300 mixed beans onto the surface of the feeding habitat (the colored tray). Be sure that the seeds are evenly scattered and not concentrated in any particular area of the habitat.

7. **STOP**! Before proceeding to the next step, formulate two hypotheses. Which predator (feeding appendage type) will have the highest fitness? Why? Which prey type (bean) will have the highest fitness? Why? Write down your two hypotheses in the appropriate spaces in the *Questions for Thought and Discussion* section at the end of this chapter. When you have formulated your hypotheses, proceed to the next step.

8. When ready, one "predator" in your group (using a stopwatch or wall clock) will announce a 60 second "feeding frenzy". During the feeding frenzy, all predators in your group will use their "beaks" to pick up as many beans as possible and transfer them to the crops. Crops may <u>not</u> be used as scoops for feeding!

9. At the end of the feeding frenzy, each predator will count the total number and types of beans (prey) captured.

10. Record the numbers in the data sheets provided.

11. Calculate fitness values for the particular predatory types and the specific prey types in your group. (Instructions for calculating fitness values are provided with the data sheets at the end of this chapter) Note: You will have to perform separate fitness calculations for the three predator types and the three prey types for a total of six calculations.

12. Clean up: Separate the beans by color, and return them to the appropriate containers in the supply area. Check the floor for dropped beans. Be sure your feeding habitats (trays) are clean.

Name: _____ Date: _____

Group Data

Habitat color for **your** predator group (circle one): Black, Tan, White

Fill in the following information:

   A. Total number of:

      1) Beans captured by plastic tweezers only: _____

      2) Beans captured by standard clothespins only: _____

      3) Beans captured by heavy duty clothespins only: _____

   B. Total number of beans captured by entire group: _____ $(A_{(1)} + A_{(2)} + A_{(3)})$

   C. Total number of:

      1) Plastic tweezers in my group: _____

      2) Standard clothespins in my group: _____

      3) Heavy duty clothespins in my group: _____

   D. Total number of all predators in my group: _____ $(C_{(1)} + C_{(2)} + C_{(3)})$

   E. Total number of:

      1) Escaped (not captured) black beans in my group: _____

      2) Escaped (not captured) tan beans in my group: _____

      3) Escaped (not captured) white beans in my group: _____

   F. Total number of all escaped (not captured) beans: _____ $(E_{(1)} + E_{(2)} + E(3))$

   G. Total number of black or tan or white beans at beginning: **100**

   H. Total number of all beans at beginning: **300**

Fitness Calculations

Using the information on the previous page, calculate the fitness values for all three predator types in your group as follows:

Plastic Tweezers fitness $= (A_{(1)} \div B) \div (C_{(1)} \div D) =$ _____

Standard clothespins fitness $= (A_{(2)} \div B) \div (C_{(2)} \div D) =$ _____

Heavy duty clothespins fitness $= (A_{(3)} \div B) \div (C_{(3)} \div D) =$ _____

Using the information on the previous page, calculate the fitness values for all three prey types in your group as follows:

Black bean fitness $= (E_{(1)} \div F) \div (100 \div 300) =$ _____

Tan bean fitness $= (E_{(2)} \div F) \div (100 \div 300) =$ _____

White bean fitness $= (E_{(3)} \div F) \div (100 \div 300) =$ _____

Predicting the Future: Applying Fitness Calculations to a Population

Suppose your initial populations of plastic tweezers, standard clothespins, and heavy duty clothespins consisted of 100 individuals each. Apply your calculated fitness values for these three predator types as follows:

100 tweezers (generation 1)   X _____ (Plastic tweezer fitness)

                                        = _____ expected # tweezers (generation2)

100 std. clothespins (generation 1)   X _____ (Std. Clothespin fitness)

                                        = _____ expected # std. clothespins (generation 2)

100 h.d. clothespins (generation1)   X _____ (H.d. Clothespin fitness)

                                        = _____ expected # h.d. clothespins (generation 2)

You can apply similar fitness calculations to your prey populations as follows:

100 black beans (generation 1) X _____(black bean fitness) = _____ black beans (generation 2)

100 tan beans (generation 1)   X _____ (tan bean fitness)   = _____ tan beans (generation 2)

100 white beans (generation 1) X _____ (white bean fitness) = _____ white beans (generation 2)

Pooled Class Data: Habitat Color and Prey (Bean) fitness

Query the other groups in your lab, and enter their prey (bean) fitness values (along with your own) in the chart below. Note: This chart was designed for a total of six groups:

	Black Bean Fitness	*Tan Bean Fitness*	*White Bean Fitness*
Black Habitat 1			
Black Habitat 2			
Tan Habitat 1			
Tan Habitat 2			
White Habitat 1			
White Habitat 2			

Questions for Thought and Discussion

1. Initial Hypotheses

   A. Before the "feeding frenzy", the initial group hypothesis regarding predator fitness was (for ALL predator types):

   B. The rationale for this hypothesis was:

   C. Before the "feeding frenzy", the initial group hypothesis regarding prey (bean) fitness was (for ALL prey types):

   D. The rationale for this hypothesis was:

2. Results

   A. Which predator in your group had the greatest actual fitness? Based on your result, did you accept or reject your hypothesis (in 1A) above? Offer detailed plausible explanations for your actual results, especially any unexpected result or rejected hypothesis.

   B. Which prey (bean) in your group had the greatest actual fitness? Based on your result, did you reject your hypothesis (in 1C) above? Offer detailed plausible explanations for your actual results, especially any unexpected result or rejected hypothesis.

3. In this exercise, simulated genetic variation within a predator community with respect to the nature of feeding appendages was observed. Similar variation in color in the prey community was also seen. In your opinion, what is the ultimate source of all genetic variation within a population?

4. After performing fitness calculations, you most likely noticed that fitness values seem centered around the number 1.

   A. What does a fitness value of 1 imply for future population growth?

   B. What does a fitness value of less than 1 imply for future population growth?

   C. What does a fitness value of greater than 1 imply for future population growth?

5. Refer to the table entitled **"Pooled Class Data: Habitat Color and Prey (Bean) fitness"**. How did the background color of **your** particular habitat influence your prey survival? Did fitness values for prey (beans) of a certain color change between groups when the background color of the habitat was different? How did habitat background color influence the ability of the predator to feed? Answer these three questions and offer explanations in complete sentences.

6. How did the background texture of your habitat influence prey survival? How did it influence the ability of the predator to feed? Answer these two questions and offer explanations in complete sentences.